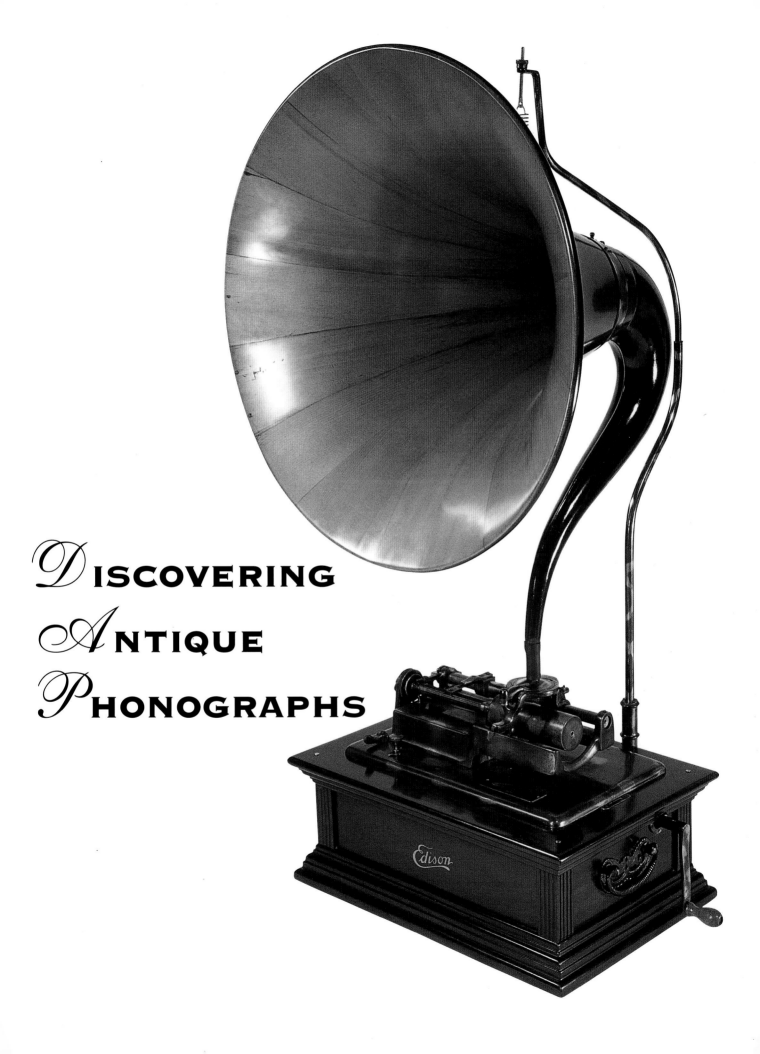

DISCOVERING ANTIQUE PHONOGRAPHS

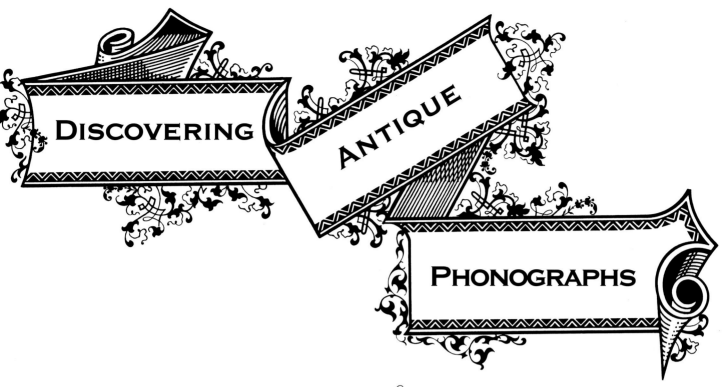

DISCOVERING ANTIQUE PHONOGRAPHS

TIMOTHY C. FABRIZIO & GEORGE F. PAUL

WITH PHOTOGRAPHS BY THE AUTHORS

Schiffer Publishing Ltd

4880 Lower Valley Road, Atglen, PA 19310 USA

DEDICATION

This book is dedicated to those who may find this book in their local bookstores or libraries and discover a welcoming door to the past.

It is curious to reflect that the Assyrians and Babylonians, 2500 years ago, chose baked clay cylinders inscribed with cuneiform characters, as their medium for perpetuating records; while this recent result of modern science, the phonograph, uses cylinders of wax for a similar purpose, but with the great and progressive difference that our wax cylinders speak for themselves, and will not have to wait dumbly for centuries to be deciphered…

With our facilities, a sovereign, a statesman, or a historian, can inscribe his words on a phonograph blank, which will then be multiplied a thousand-fold; each multiple copy will repeat the sounds of his voice thousands of times; and so, by reserving the copies and using them in relays, his utterance can be transmitted to posterity, centuries afterwards, as freshly and forcibly as if those later generations heard his living accents. Instrumental and vocal music—solos, duets, quartets, quintets, etc.—can be recorded on the perfected phonograph with startling completeness and precision.

How interesting it will be for future generations to learn from the phonograph exactly how Rubinstein played a composition on the piano, and what a priceless possession it would have been to us, could we have Gen. Grant's memorable words, 'Let us have peace,' inscribed on the phonograph for perpetual reproduction in his own intonations!

-Thomas A. Edison, writing in the *North American Review,* June 1888.

Library of Congress Cataloging-in-Publication Data

Fabrizio, Timothy C.
Discovering antique phonographs / Timothy C. Fabrizio & George F. Paul; with photographs by the authors.
p. cm.
Includes bibliographical references.
ISBN 0-7643-1048-8
1. Phonograph--Collectors and collecting. 2. Phonograph--History. I. Paul, George F.
II. Title.
TS2301.P3 F299 2000
621.389'33'075--dc21
99-054435

Designed by Bonnie M. Hensley
Type set in Milano LET/Souvenir Lt BT

ISBN: 0-7643-1048-8
Printed in China
1 2 3 4

Published by Schiffer Publishing Ltd.
4880 Lower Valley Road
Atglen, PA 19310
Phone: (610) 593-1777; Fax: (610) 593-2002
E-mail: Schifferbk@aol.com
Please visit our web site catalog at **www.schifferbooks.com**

In Europe, Schiffer books are distributed by Bushwood Books
6 Marksbury Avenue Kew Gardens
Surrey TW9 4JF England
Phone: 44 (0)208-392-8585; Fax: 44 (0)208-392-9876
E-mail: Bushwd@aol.com
Free postage in the UK. Europe: air mail at cost.

This book may be purchased from the publisher.
Include $3.95 for shipping. Please try your bookstore first.
We are interested in hearing from authors with book ideas on related subjects.
You may write for a free printed catalog.

Contents

Acknowledgements

The authors would like to thank those collectors and institutions whose items appear anonymously, and under the following attributions: Jean-Paul Agnard, Gary W. Blizzard, Roger Bodenheimer, Martin F. Bryan, Lou Caruso, R. Chase, Domenic DiBernardo, Doug Defeis, the John and Nancy Duffy family, David Giovannoni, Howard Hazelcorn, Charles Hummel, Allen Koenigsberg, Dan and Sandy Krygier, Charles McCarn, William G. Meyer, Alan H. Mueller, Walter and Carol Myers, Ray Phillips, Michael and Suzanne Raisman, Robin and Joan Rolfs, Sam Saccente, Steve and Ellie Saccente, Jasper Sanfilippo, Lawrence A. Schlick, Sam Sheena, Michael Sims, Bob and Karyn Sitter, Steve and Theresa Sposato, Donald Walls, David Werchen, and John Woodward. We are especially indebted to the National Park Service, Department of the Interior, Edison National Historic Site, and the Herculean efforts of staff member Jerry Fabris for providing access to some of the rarest treasures in the Edison legacy.

We would also like to extend our gratitude to: *apprenti sorcier* Jean-Paul Agnard, the punny Bob Amos, Julien Anton, Lynn Bilton, Phil Carli, Bob Carver, the indispensable W.C. Chin, Dom Coladonato, George Cook, Aaron Cramer, Kate and Denise Fabrizio, Cousin Bob Fenimore, the imaginative Ryan Fraser, the Puckish George Glastris, Tim Gracyk, Dave Hasler, the entire Heffernan Family, our brilliant designer Bonnie M. Hensley, our super-deluxe editor Molly Higgins, the hopelessly clever Howard Hope, Garry "the Gremlin" James, Larry Karp, Charlotte Keating, the urbane Philippe Le Ray, Larry Levitz for his invaluable help with Keen-O-Phone, Peter "the Great" Liebert, Rob "High Tech" Lomas, John McBroom, the witty and irreverent John T. Mayo, John D. Miller who generously supplied an important original document, Dr. Donald Munger, the ever-enthusiastic Barbara, Jess and Tim Paul, the always-encouraging Joan Paul, the late Dr. John E. Paul, Joe Pengelly, in memoriam to the gracious Nancy Phillips, the redoubtable Robert Ridgeway, Yves Rouchaleau, Zach Saccente, Rick Shiano, the lovable Norm and hug-able Janyne Smith (or is it the other way around?), Marie-Claude Stéger *"La Reine du Phonographe,"* and, as always, the one-and-only Ray Wile. A very special note of appreciation must be directed to the very dedicated scholars John Woodward, Christian Eric, Marc Christian and Mike Khanchalian whose *ongoing* research, re-recording and restoration of the earliest Stroh apparatus illustrated here was indispensable to us.

Finally, thank you again to the forever-ebullient Peter Schiffer, the masterful and commanding Doug Congdon-Martin and all the friendly and cooperative folks at Schiffer Publishing—and where, please, does Peter buy his optimism pills?

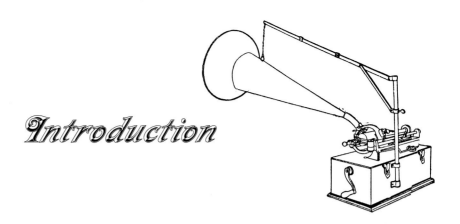

Introduction

With the Third Millennium upon us, public attention has suddenly focused on the fruits of the past century. For fanciers of mechanical music, it is surprising to consider that much of the most important talking machine development occurred during the century *before* the last. It is time to look with new eyes upon the earliest landmarks in the progress of recorded sound. The reader is about to discover a breathtaking array of incunabula. Through our illustrations and captions, we hope to inspire further thought and discourse about the people and events which shaped the history of the phonograph throughout the "acoustic" (pre-electric reproduction) era, 1877-1929.

When we wrote the introduction to our first book, *The Talking Machine, An Illustrated Compendium 1877-1929* (Schiffer, 1997), we lamented the fact that the phonograph was given little attention in museums and institutions in the United States. What could be more common to our *everyday* lives than recorded sound? Yet, the phonograph has been treated as a kind of country cousin of the telephone, electric light, automobile and airplane. We wish to commend the Edison National Historic Site for making a dramatic step toward helping the phonograph to achieve parity among the great inventions of humanity. The Site kindly permitted us to photograph and document some of the rarest objects in Thomas Edison's sound recording legacy, many of which would otherwise not be known to the public.

In our first book, we likened the antique phonograph to a time machine. Thousands of collectors and aficionados make regular visits to the past, awash in the sound waves of a vanished society. To close one's eyes and travel to the Milan hotel room where, in 1902, Caruso is making his first Gramophone records; or to the London cellar where, in 1899, the Hotel Cecil Orchestra is playing; or to "somewhere in France," where General Pershing is exhorting the Allies on to victory; or to thousands of other places and times around the world, is to witness history in a way not possible before 1877. This book is a study of the marvelous talking machines which transport us to the past, and simultaneously enhance the present.

The *visual* study of an *acoustic* device might seem incongruous at first, but it is the first step toward seeing beyond the gleaming nickel plating, glowing wooden cabinets, and cheerful flower horns to something deeper. We hope to demonstrate how the talking machine evolved during its first half-century, from a novelty which merely reflected existing societal mores, to an industrial force capable of inciting social change. The talking machine itself greatly influenced our history and, in refined and various forms, influences our present lives.

In the years following the publication of our books, *The Talking Machine Compendium* and *Antique Phonograph Gadgets, Gizmos and Gimmicks* (Schiffer, 1999), we continued traveling, photographing and researching. We came to realize the stunning variety of antique phonographs remaining to be discovered and documented in collections around the world. Out of these further explorations, this book came to life. Through the over 400 entirely new color illustrations contained here, the reader will follow the progress of the acoustic talking machine from its crude beginnings to its most sophisticated heights. Whether you have read our previous books or not, prepare to embark on an astounding journey, and discover the diversity and endearing charm of the antique phonograph.

CHAPTER ONE
1877-1893

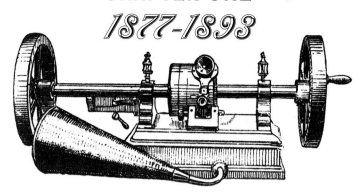

The precipitate developments in sound experimentation during the year 1877, which culminated in December with the completion of the first Phonograph, had germinated some twenty years earlier. It is ironic that the sudden inspiration that elevated long-languishing ideas to a higher plane should have imbued two visionaries from vastly different worlds almost simultaneously.

To find the wellsprings of the Phonograph, we must begin with a Frenchman, Léon Scott de Martinville (1817-1879). A printer by trade, Scott experimented with sound, and in 1857 invented a device for visually depicting sound waves. Scott called his apparatus the Phonautograph. This machine, as improved in 1859, used a drum blackened with carbon to show the tracings of a bristle, or stylus, which was attached to a stationary diaphragm. As the drum was rotated beneath the stylus, sound directed at the diaphragm would appear as an undulating line on the surface of the cylinder. Although Scott's invention became part of the equipment of scientific institutions and universities around the world, decades would pass before the next step toward realizing the dream of recorded sound would be taken.

Early in 1877, another Frenchman, Charles Cros (1842-1888), conceived of adapting the Phonautograph so that the sound waves drawn by it might be aurally reproduced. Cros suggested using a photoengraving process to make the "recorded" undulations permanent. He reasoned that passing the Phonautograph stylus over engraved sound depictions would cause the diaphragm to replay the sound. The merit of his theory, however, was destined to be investigated by others. Cros, lacking the resources to support a laboratory, never learned whether his idea would work. He did, however, describe his proposed invention (the Paleophone) in a scientific paper that he deposited with the French Academy of Sciences on April 30, 1877. This document was sealed, and remained unread for a number of months, during which time another man catapulted theory into action.

In the United States during 1877, the prolific inventor Thomas A. Edison was investigating improvements to both the telegraph and the telephone. Chance would inspire him to investigate sound recording. In one story, a wire connected to a vibrating telephone diaphragm pricked Edison's finger, causing him suddenly to imag-

1-1. Edouard-Léon Scott de Martinville (1817-1879), whose profession as a printer suggested to him a way that sound could be visually displayed, invented the disc Phonautograph in 1857 and a cylinder version (shown) in 1859. The Phonautograph depicted the physical characteristics of sound as lines transcribed on a carbon-covered drum. Two instrumental figures in the history of recorded sound were greatly influenced by Scott's device. In 1877, Frenchman Charles Cros developed a theoretical application of Scott's principles by which recorded sound might be replayed. Cros never constructed his proposed apparatus, the Paleophone. However, his method of transforming the Phonautograph's transcribed sound waves into a three-dimensional record, through a photoengraving process, inspired Emile Berliner to delve further. By the late 1880s, Berliner had already tested the practicality of Cros' ideas, and begun to develop his own approach to sound recording. In Berliner's process, a recording stylus removed material from a wax-coated zinc disc, which was then acid-etched. Berliner's refinement of the theory behind the Phonautograph resulted in the first commercially feasible disc recording method. *Courtesy of Allen Koenigsberg.* (Value code: VR)

ine how such a wire stylus might leave a permanent impression of sound waves in a pliable medium. Another story had Edison working with his Embossing Translating Telegraph. This invention recorded telegraphic signals as indentations on a wax-coated paper disc. To repeat the message, a mechanical finger would follow the indentations and transform them back into electronic impulses. As the machine indented the rotating disc, it made a humming noise, which seemed to Mr. Edison like murmured speech. The inventor was witnessing, in fact, the harbinger of the Phonograph.

1-2. Edison's Embossing and Translating Telegraph was an immediate precursor of the "Tinfoil Phonograph." Both devices embossed, or indented, pliable material to create a "record." The Embossing Telegraph relayed telegraphic messages by indenting a wax-covered paper disc. When Edison envisioned a disc-playing version of the "Tinfoil Phonograph," which can be seen in his laboratory notes and his British patent (No.1644, 1878), he virtually copied the design of the Embossing Telegraph. Surviving documents make it clear that Edison constructed at least one prototype "Disc Tinfoil" device, though none have come to light. *Courtesy of the Edison National Historic Site.* (Value code: VR)

EDISON'S PHONOGRAPH IS BORN

Edison and his laboratory staff started working in earnest on a recorded-sound device on July 18, 1877. By early September, the apparatus had been named "The Phonograph," but a public announcement was postponed as experimentation went forward. The Novem-

ber 17 issue of *Scientific American* printed a basic description of the Phonograph, even though Edison had yet to arrive at a successful prototype. It was December 6, 1877 when Edison's assistants finished the machine by which sound would be recorded and replayed for the first time. According to Edison's recollection, he spoke some inconsequential words into it (usually described as the nursery rhyme "Mary Had a Little Lamb"), and it distinctly repeated them. Just three days earlier, Charles Cros had ordered his secret paper on the Paleophone opened by the Academy of Sciences in Paris. Although he had already made his theory public (in the periodical *La Semaine du Clergé*, October 10, 1877) the unfortunate months of delay had doomed the Frenchman to eternally wander that circle of Scientific Hell reserved for dreamers.

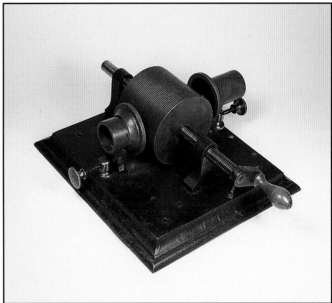

1-3. After overhearing the murmuring of the Translating Telegraph, Edison's epiphany set into motion ideas that would culminate, on December 6, 1877, in the invention of the Phonograph. This achievement followed weeks of increasingly feverish experimentation. Finally, Edison's assistant, John Kruesi, constructed the prototype shown here, and presented it to Edison for his approval. Edison reputedly spoke a nursery rhyme, *Mary Had a Little Lamb,* into the machine, which then repeated the words recognizably. *Courtesy of the Edison National Historic Site.* (Value code: VR)

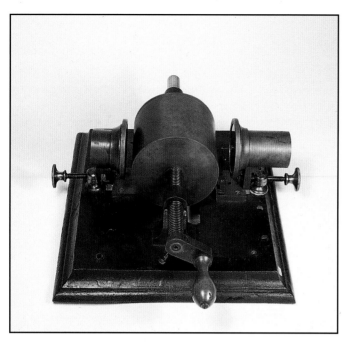

1-4. To make a recording, the drum or mandrel of the first Phonograph was covered with a thin sheet of tin or lead foil. The helical groove in the mandrel allowed the foil to be indented by the recording stylus. The machine had one recording (left) and one reproducing (right) mouthpiece. A steel stylus pressing against a diaphragm was used in both recording and reproducing. Apart from the wooden winding handle, the first Phonograph was made entirely from metal. Despite its phenomenal impact on the public, Edison's Phonograph would languish for years without significant development. *Courtesy of the Edison National Historic Site.* (Value code: VR)

Edison's successful design for the Phonograph functioned in a manner not unlike the Embossing Translating Telegraph that had provoked Edison's quest to record sound. Although it is clear from Edison's British patent for the Phonograph, granted in 1878, that the inventor had considered recording sound on flat discs, the format he selected was a cylinder. A mechanical arrangement similar to the Phonautograph was used to pass a rotating drum beneath a stylus connected to a stationary diaphragm. However, Edison's drum, or mandrel, was cut with a spiral groove. To make a record, a thin sheet of tin or lead foil was wrapped around the mandrel. As this flexible covering passed beneath the stylus the recessed groove in the mandrel allowed the foil to *indent*, creating a sound recording. Reversing the process reproduced the sound.

Though the Phonograph was celebrated far and wide as yet another miracle wrought by the genius of Mr. Edison, the sizzle fizzled fast. Within a year the Phonograph had entered a kind of stasis from which it would emerge only after rival experimenters took the initiative. The Edison Speaking Phonograph Company, formed to sell the devices, failed to find a market in the United States. Extant machines suggest, however, that in Europe the influence of the "Tinfoil Phonograph" was more far-reaching. To over-simplify national traits, the American personality tended to rapidly embrace and abandon fads, whereas the European mind was more pragmatic.

In Europe, especially France and Germany, "Tinfoil Phonographs" were produced in greater variety, sold in larger numbers and remained popular for a longer period of time than in the United States. In Great Britain, the London Stereoscopic (and Phonographic) Company was granted the right in March 1878 to manufacture versions of Edison's "Tinfoil Phonograph." This British firm would offer its phonographs, of which a weight-driven model is the most commonly recognized today, until the mid-1880s. John Matthias Augustus Stroh (called "Augustus," 1828-1914) claimed to have produced the first Phonograph constructed in Great Britain. To quote an article by T.C. Hepworth, in the March 4, 1903 issue of *Cassell's Popular Science*:

> But soon after its [the Phonograph's] *début* in America a gentleman who had seen and heard it came over here [England] and gave a verbal description of it to Mr. Augustus Stroh, and that clever worker was able to produce within a very few days a machine which would talk....This early phonograph...was shown [February 1, 1878] at the Royal Institution [originally, the Society of Telegraph Engineers] during a lecture by Sir W.H. Preece, on which occasion the late Poet-

1-5, A *carte de visite* showing a young and rather dour Edison soon after his invention of the Phonograph. *Courtesy of Sam Sheena.*

Laureate, then known simply as Alfred Tennyson, occupied the chair. After the mechanism of the instrument had been explained, Professor Tyndall, in compliment to the poet, shouted into it 'Come into the garden, Maud.' Presently the phonograph repeated the words in a comical falsetto, and the effect upon the audience was so startling that they broke into tumultuous applause…

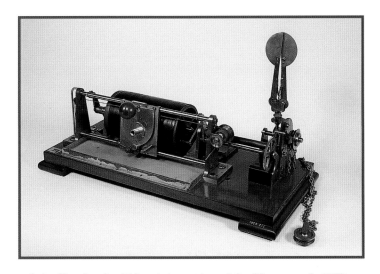

1-6. Shortly after Edison's invention of the Phonograph, William H. Preece commissioned Augustus Stroh to build a "Tinfoil Phonograph" that closely followed the Edison pattern. Stroh quickly completed the task, then designed and built the far more sophisticated model shown here, demonstrating it in February 1878. Unlike the hand-driven models, this machine was equipped with a weight-driven motor, the world's first motor-powered talking machine. The air governor, with its hemispherically shaped vanes, would spin as the machine operated. Stroh dispensed with the laterally traveling mandrel of other "Tinfoil Phonographs," substituting a finely machined traversing reproducer/recorder driven by a fusee chain. Although Stroh's February 1878 demonstration of this machine employed tinfoil recordings, the mandrel was shortly thereafter equipped with a permanent waxen surface. Advanced as this machine was for its time, it is the recording which is perhaps most interesting. Embossed into the wax at 140 lines-per-inch, and playing at 40-45 rpm, are recordings of a brass ensemble, a woman singing, a cornet solo, and a man addressing an audience (see inset). No documentation specific to these recordings has been found, but the machine itself provides a clue. The diaphragm is of a design abandoned by Stroh by early 1879, when he discovered that "…a stretched membrane of thin india-rubber rendered rigid by a cone of paper, was found to give the best effects." The purpose of this machine was clearly to maximize recording and playback quality, so it seems that if the recording dated from later than early 1879, the machine would have been equipped with the improved diaphragm. Additionally, the content of the recording is typical of that which would have been demonstrated in the very earliest days of the "Tinfoil Phonograph." By mid-1878, Stroh's attentions had been directed to the analysis of the human voice using phonographic apparatus, probably relegating this pioneering demonstration machine to a back room. Although currently impossible to determine conclusively, this machine may carry the world's earliest playable recording. Base measurements are 10 3/4" x 16 1/2". Mandrel dimensions are 7 1/2" long x 4 1/4" diameter. *Courtesy of John Woodward.* (Value code: VR)

AUGUSTUS STROH, SCIENTIST OF SOUND

Mr. Stroh's life and work bear noting here, since they were intertwined with the history of recorded sound. Stroh was German by birth, and was apprenticed at a young age to a clock-maker. His mechanical aptitude became immediately apparent, and he received first place in a clock-making competition. Traveling to London in 1851 to view the Great Exhibition in Hyde Park, he was so excited by the quality of British science that he emigrated and became a naturalized British subject. Stroh entered into a partnership (until 1875) with Sir Charles Wheatstone to design and produce various apparatus related to telegraphy and acoustics. As already mentioned, he constructed a "Tinfoil Phonograph" early in 1878, and at least one other example soon after. Stroh's second version of the "Tinfoil Phonograph" was demonstrated before the Society of Telegraph Engineers on February 27, 1878. It was weight-driven, and highly sophisticated in comparison to the crude and clunky mechanics of most contemporary phonographs (see illustration 1-6).

The "Tinfoil Phonograph" that Stroh exhibited in London was described by Mr. Hepworth as follows:

> He [Stroh] saw that the instrument [Edison's original Phonograph] was open to great improvement, and his first step was to give its cylinder a more steady motion than is possible by a wheel turned by hand. He therefore fitted it with a clock-work train driven by a weight….Not only, we are told, was the articulation of spoken words far more perfect, but songs sung into the instrument by Sir W.H. Preece and others were 'reproduced with very respectable correctness.'

The following is a transcription of the spoken portion of the Stroh wax recording:

> *Ahem…Well, gentlemen, I'm really much obliged to you all for playing for me this evening, and I think what you have done for me is all very well indeed. I wish you a happy new year and a very good night to you all. Good night. [coughs comically] I hope you haven't got such a bad cold as I have!*

It is interesting to note that the musical portion of the recording contained no excerpts from Gilbert & Sullivan. By the mid-1880s, the infectious melodies from Gilbert & Sullivan operettas had permanently entered British musical repertoire. The musical content of the Stroh cylinder more resembles the work of 1870s composers such as Jacques Offenbach. It is tempting to consider that this "happy new year" might have been uttered as early as January 1879.

In 1878, following his initial experimentation with the Phonograph, Stroh revived the controversial idea originated by Charles Cros the previous year: that an accurate, replayable record of sound could be derived from the tracings of the Phonautograph. Stroh experimented with and refined the Phonautograph as part of his analysis of sound, but his focus was not on sound reproduction. Rather, his purpose was an innovative study of the human voice. In order to understand what Stroh was attempting to demonstrate, some very basic background is necessary.

Hermann Helmholtz (1821-1894), regarded as the "Father of Acoustics," had postulated in 1862 that a particular human voice was most readily distinguishable from other voices through its production of vowel sounds. It was subsequently confirmed that this is due to widely variable arrangements in the components or tones of human vowel sounds. These components may be more easily understood in the analogy of a musical chord. Each vowel (chord) is made up of several tones. The lowest tone is called a fundamental or "prime." Additional tones or harmonics are present at higher frequencies, as many as 18 or 20 in the human voice. These harmonics or "partials," like notes in a chord, can be altered somewhat by the speaker to make different vowel sounds while retaining the same fundamental frequency.

The sound of the vowel is dependent upon both the fundamental frequency and the particular arrangement of the harmonics or partials. Changing either will change the sound of the vowel. Stroh, along with William Henry Preece (chief engineer of the General Post Office of Great Britain), sought to demonstrate this acoustic principle mechanically. During 1878, Stroh constructed four separate machines whose combined goal was to "synthesize" a human voice. The use of the word "synthesize" can be rather misleading here. To modern-day ears, the parlance suggests *artificial* speech not derived from a human source, such as Faber's automaton of the human speech organs, which replicated lips, windpipe and larynx all operated by pedals and buttons. Stroh, however, sought to combine human and mechanically-generated voice components. The "synthetic" coinage referred to the possibility that through his apparatus, the parts of one individual's speech might be combined with another's, or altered mechanically to produce a "synthetic" voice.

Stroh's first machine was called "the synthetic curve machine." Like the Phonautograph, it traced sound waves on paper. However, Stroh's device used a mathematically-formulated relationship of gears to depict on paper sound waves similar to those generated by the human voice. Stroh's second machine transferred the pattern of curves produced by the first machine onto the circumference of a brass disc. Each disc contained the sound of a single harmonic or partial. The third machine was essentially a phonograph, in that it could

play one or more of these brass discs in any combination: *synthetically* creating and modifying a vowel. The fourth machine, called an "automatic phonograph," played brass discs which, unlike those used with the third machine, incorporated compound partials: in other words, multiple tones. By sliding the diaphragm from disc to disc, the various combinations of sound could be reproduced "automatically." It was this fourth machine with which Preece and Stroh accompanied a paper before the Society of Telegraph Engineers on February 27, 1879 entitled *Studies in Acoustics: On the Synthetic Examination of Vowel Sounds.*

The ninety-eight existing Stroh brass discs present intriguing evidence from the dawn of human voice recording. In addition to vowel components, several discs contain consonants, and even the words "Mama" and "Papa." These words could only have resulted from a human speaker; Stroh's machinery was incapable of artificially replicating entire words. In light of this, Stroh's apparatus takes on a new dimension. It was not merely

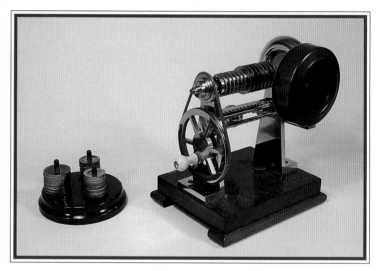

1-7. As Augustus Stroh and William H. Preece proceeded into a study of the components of vowel sounds, their experiments resulted in the construction of four different machines in 1878. The first machine traced mathematically-designed waveforms on paper. The second machine transferred these artificial or synthetic waveforms onto the periphery of a brass disc. The third machine could play up to 8 of these discs in any combination, melding the different waveforms or frequencies into a vowel sound. Each disc carried a recording of approximately 1/2 to 1 second long. The fourth machine (shown) was called the "automatic phonograph." The discs played by this machine incorporated *multiple* waveforms in various combinations. By sliding the reproducer from disc to disc, these various combinations of sound frequencies could be compared. This attempt to synthetically create vowel sounds met with partial success, and the results were presented by Preece and Stroh to the Society of Telegraph Engineers on February 27, 1879. Comparisons to a genuine human voice were inevitable, and some of the existing 98 brass discs contain recordings of actual voices (including the words "mama" and "papa."). The actual dates of these recordings are difficult to pinpoint, but one of these discs may contain the world's oldest playable record. *Courtesy of John Woodward.* (Value code: VR)

a vehicle for theoretical study. It actually *spoke*, and it spoke in the voice of a once-living individual. These little brass discs might even contain the earliest known recording of human speech.

It is interesting to note that Stroh's work was more a vindication of Charles Cros than the experiments of Emile Berliner, some ten years later. Although Berliner's investigation of sound recording is cited nowadays as support for Cros' theories, Berliner was merely inspired by the Phonautograph, and Cros' hypothetical modification of it (the Paleophone). Berliner ultimately failed to satisfactorily reproduce sound by Cros' method of converting a visible tracing into a permanent record.

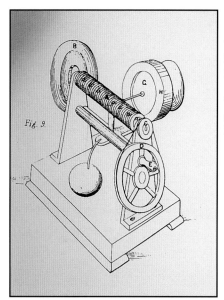

1-9. An illustration of the Stroh "automatic phonograph" as it appeared in the February 1879 proceedings of the Society of Telegraph Engineers in London.

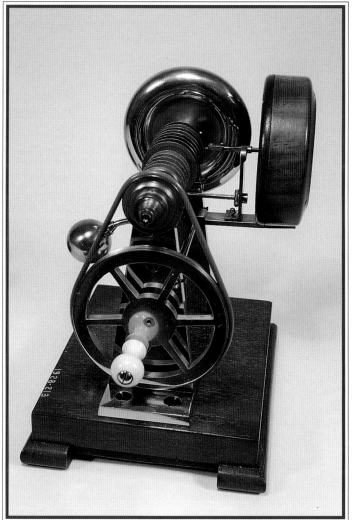

OTHER INNOVATORS IN THE FIELD

On January 7, 1878, representatives of the Ansonia Clock Company of New York City received a license from Edison to construct a talking clock. After a period of time the project was put in the hands of a mechanic named Frank Lambert. Lambert ultimately produced two prototype phonograph mechanisms (to be used with the proposed clock), one of which survives today. The existing device has a lead-sheathed mandrel on which a recording of the hours is still playable. Scholars have assumed that Lambert's was the oldest extant "playable"

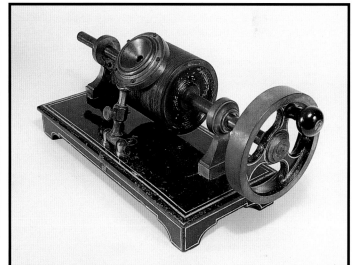

1-8. Another view of the Stroh "automatic phonograph." The stylus and its linkage to the diaphragm can be clearly seen, as well as the knurled adjustment screw. Base dimensions are 5 1/2" x 7". Note that some of the discs were larger in diameter than others. The larger size increased the relative surface speed of the disc, allowing the intricate undulations to be captured and reproduced with more detail and greater fidelity. Twenty years later, the "Graphophone Grand" would employ the same principle. *Courtesy of John Woodward.* (Value code: VR)

1-10. A "Tinfoil Phonograph" circa 1879 from the shop of a precision instrument maker. The machine is marked: "J. Niora, Torino [Italy]." The black gilt-decorated base measures 5 3/4" x 12 1/2". One of the features soon added to "Tinfoil Phonographs" was a retractable feednut, as seen on the left-hand mandrel shaft support. When the nut was disengaged, the mandrel assembly could be quickly returned to the starting position. (Value code: VR)

recording. The date of the Ansonia-Edison contract has been established, but it is unknown when the particular surviving talking clock phonograph and its lead recording were created. Lambert's own testimony in a subsequent court case indicated that after he was engaged, a number of months elapsed before he constructed his first machine. We know that by mid-February 1879 Stroh and Preece's work on the "automatic phonograph" was fully completed. There is not enough evidence to determine whether "mama" and "papa" spoken by Stroh's "automatic phonograph" are contemporary with the hours spoken by Lambert's phonograph mechanism. It is certainly possible.

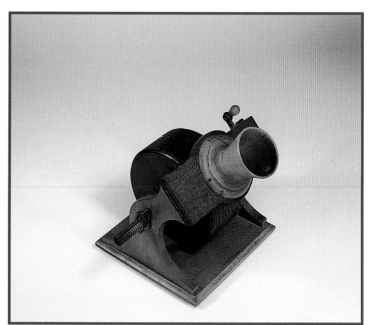

1-13. A small French "Tinfoil Phonograph" of 1878 or 1879, made primarily of wood. The mandrel is plaster, coated with a dark red sealant. The machine is marked "D. Vital" and measures 5 3/4" square. "Professor and Lecturer" Monsieur Vital published a 22-page thesis on "Tinfoil Phonographs" in 1879, with illustrations of this machine. (Value code: A)

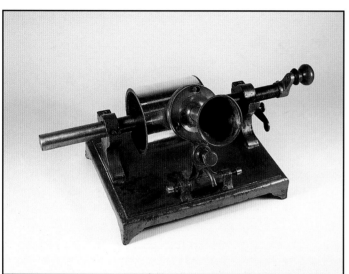

1-11. A French "Tinfoil Phonograph" by Urban Marie Fondain, 12 Rue Richer, Paris, circa 1880. Monsieur Fondain was a pioneer in the French "Tinfoil Phonograph" trade, which outlasted him (he declared bankruptcy in April 1881). Eugene Ducretet acquired some of Fondain's inventory, and constructed apparatus of a similar design. The mandrel measures 3 1/2" diameter. *Courtesy of Jean-Paul Agnard.* (Value code: A)

1-14. A very large, unmarked "Tinfoil Phonograph," probably pre-1880. The brass mandrel measures 5 3/4" diameter x 5 7/8" length. *Courtesy of Jean-Paul Agnard.* (Value code: VR)

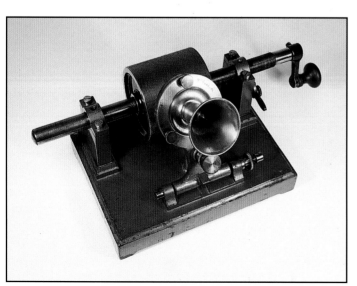

1-12. A Ducretet "Tinfoil Phonograph" from about 1880, marked "Ducretet et LeJeune, Paris, 7980-25." In its general shape and the execution of its parts, especially the bronzed iron base, it very much resembled the work of Fondain, from whom Ducretet drew both inspiration and spare parts. Not surprisingly, the mandrel measures 3 1/2" diameter. *Courtesy of the Domenic DiBernardo collection.* (Value code: A)

1-15. A delicate, colorfully-decorated European "Tinfoil Phonograph," circa 1880. In Europe, the red-on-black motif was a popular form of talking machine decoration, especially in Germany. Machines such as this were intended merely to demonstrate the principal of sound recording and did not incorporate the heavy flywheels or large diameter mandrels necessary to conduct public exhibitions or serious scientific study. The tiny mandrel measures 2 3/4" diameter. (Value code: B)

1-16. A London Stereoscopic Company "Tinfoil Phonograph," equipped with a spring-driven motor. Note that the governor vanes here are round. Base measures 9 3/4" x 21 3/4". (Value code: VR)

1-17. The motor of the previous example in operation, showing the governor vanes fully extended. (Value code: VR)

1-18. This unmarked spring-driven "Tinfoil Phonograph" is a direct adaptation of a hand-driven model, with a base wide enough to accommodate the motor. The mahogany base measures 11 1/2" x 23 1/2". (Value code: VR)

1-19. A London Stereoscopic Company hand-driven "Tinfoil Phonograph" with fly-wheel. This appears to be the model listed in the company's 1886 catalogue as selling for £10 10 0. Although measuring only 9 1/2" x 14 3/4", this massive cast-iron machine would be too heavy for most people to lift without help. *Courtesy of Ray Phillips.* (Value code: VR)

1-20. A London Stereoscopic Company "Tinfoil Phonograph" with a weight-driven motor. Note the upright vanes of the air-governor to the far right. This British firm was licensed by Edison in 1878 to exploit his invention in Great Britain. Although it never attempted to tap the home entertainment market, the London Stereoscopic Company continued to offer "Tinfoil Phonographs" for scientific purposes, and listed three different models as late as 1886. At that time, the version shown sold for £25. Measures 49 3/4" from floor to top of table, 29 1/2" wide. (Value code: VR)

1-21. A "Tinfoil Phonograph" of unknown origin. It incorporated two features not usually seen in a "Tinfoil" apparatus. Firstly, the reproducer was mounted on a moveable carriage driven by a feedscrew. Secondly, a quick-disengage mechanism, controlling the half-nut bar, permitted an easy return of the carriage to the starting point. This could be either a late example of "Tinfoil" experimentation (circa 1890), or a work of remarkable vision from an earlier period. *Courtesy of the Edison National Historic Site.* (Value code: VR)

1-22. This unmarked "Tinfoil Phonograph" strongly resembles an Edison Speaking Phonograph Company "Parlor" model. The mandrel is coated with wax, and to the left is a nozzle for directing a jet of air at the modulations recorded in the wax. During their experiments with pneumatic reproduction of sound in the early 1880s, the Volta Laboratory Associates (Alexander Graham Bell, Chichester Bell and Charles Sumner Tainter) tested a similar apparatus. It is tempting to think that the instrument pictured is a survivor of that period. (Value code: VR)

Like Stroh, Alexander Graham Bell was a dedicated professional scientist. Having received an award of money from the French government in 1880, Bell formed the Volta Laboratory in Washington, D.C., to investigate recorded sound. Working with his cousin, Chichester Bell, and a model-maker named Charles Sumner Tainter, Bell explored many aspects of recording and reproduction. The greatest success of the Volta Associates, however, was their development of incised recording in a wax medium. The team began simply: by filling the helical groove in the mandrel of a "Tinfoil Phonograph" with wax. Here, at least, was the opportunity to *replay* a recording. Two major deficiencies of the "Tinfoil Phonograph" were simultaneously overcome. Not only was tinfoil a poor recording medium; but tin record sheets, once removed from the machine, could never be re-aligned with sufficient accuracy to be replayed.

The Volta Associates continued to work over the next five years, and eventually developed a wax-coated cardboard cylinder record that could be removed or replaced at will, without damage to the recorded surface. However, by this point, it became clear that what the experimenters had achieved was not merely an improvement of the Edison Phonograph, but an entirely new device. They lightheartedly dubbed their creation the "Graphophone." What began as a play-on-words eventually became a fierce and formidable rival of Edison's Phonograph.

On January 6, 1886, the Volta Graphophone Company was formed. From this seed would grow the American Graphophone Company, manufacturer of a treadle-powered talking machine, which used removable, replayable wax-coated cylinder records and interchangeable, gravity-pressure floating heads for reproducing and recording. Edison was stunned; he had let the development of recorded sound get away from him. As soon as he learned of the Graphophone breakthroughs, he began plans to improve his Phonograph.

EDISON'S ENTHUSIASM REAWAKENS

During the latter part of 1886, Ezra T. Gilliland, an engineer of Edison's acquaintance, was engaged to develop new ideas for the Phonograph. Subsequently, in November 1887, Edison sent a wooden model of an electrically powered Phonograph to Gilliland's workshop in Bloomfield, New Jersey. The new design represented a huge step forward. Gilliland constructed a working model of this machine, which featured a reproducer and a recorder held in an easily pivoted double carriage assembly (the so-called "Spectacle" attachment), and a removable cylinder record made of solid wax (actually a mixture of fats and waxes). This "New" Phonograph, which was described and illustrated in the December 31,

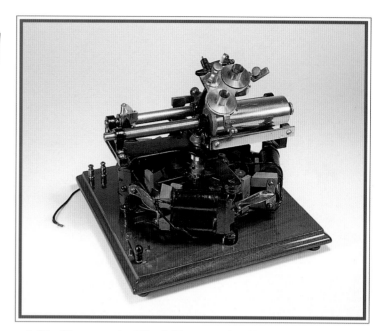

1-24. We see in the "New" Phonograph inklings of what was to become the "Perfected" or Edison Class "M" Phonograph. Though direct friction drive of the mandrel was soon abandoned, the vertical motor alignment, 100 threads-per-inch feedscrew, 4" long mandrel, and pivoted reproducer/recorder carriage were all elements that would reappear in future models. *Courtesy of the Edison National Historic Site.* (Value code: VR)

1-23. In 1887, Edison's research staff began a concerted effort to re-configure the Phonograph. The groundbreaking work of Alexander Graham Bell's Volta Laboratory had come to Edison's attention. It was evident that, unless he acted quickly, others would forever deprive him of a place in any foreseeable talking machine industry. In his Bloomfield, New Jersey workshop, Edison's associate, Ezra T. Gilliland, constructed a working model of an entirely new type of Phonograph from specifications sent to him by Edison. This so-called "New" Phonograph (shown) was completed in November 1887, and employed electric motor drive to play interchangeable wax cylinders. This machine is pictured in its mahogany carrying case. *Courtesy of the Edison National Historic Site.* (Value code: VR)

1-25. In 1888, after Ezra T. Gilliland completed work on the prototypes of the "New" Phonograph, Edison put him in charge of Phonograph research and development. Furthermore, Gilliland was made sole sales agent for the rapidly evolving Phonograph. In Edison's enthusiasm to bring a practical Phonograph to the market, the exclusive sales contract made in 1878 with the Edison Speaking Phonograph Company was conveniently forgotten. The instrument shown is from May 1888, and embodied various Gilliland refinements of the "New" Phonograph. The friction drive was moved from the mandrel to the left end of the main shaft. In addition, the motor was suspended from a cast, decorated bedplate, and entirely enclosed in a wooden cabinet. Note the attribution to Gilliland's sales agency, "The Edison Phonograph Co., New York, U.S.A." Before any of the Gilliland "Improved" Phonographs could be marketed, Edison stepped in and advanced the design even further. During a 72-hour period of ceaseless labor in mid-June of 1888, Edison and his associates created what would come to be called the "Perfected" Phonograph, the immediate predecessor of the Class "M." *Courtesy of the Edison National Historic Site.* (Value code: VR)

1887 issue of *Scientific American*, led Edison to order from Gilliland a small number of the redesigned Phonographs, mounted in portable wooden cabinets that enclosed and protected their electric motors. These became known as the Gilliland "Improved" Phonographs.

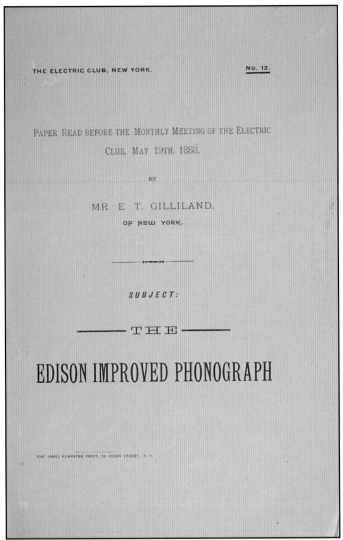

1-26. Removing the motor access plate of the previous example reveals some surprising inscriptions. The date "21/7/88" may be seen to the right. Additionally, there are five sets of initials and two names scratched into the paint. From top to bottom, these are: "JHS" (James H. Saville, vice-president and general manager of the American Graphophone Company), "C.J. Bell" (president of the American Graphophone Company), indecipherable, "Chas. S. Tainter" (Charles Sumner Tainter, principal patentee for American Graphophone), "CW" (Cleveland Walcutt, head bookkeeper and cashier of the North American Phonograph Company), "E.M.N." (unknown), and "W.G.D." (unknown). The meaning behind these markings is made clear through a receipt dated July 21, 1888, signed by James H. Saville: "Received of Clevelend Walcott [sic], sent by Jesse H. Lippincott from New York City, to deliver to this company [American Graphophone] two of Mr. T. A. Edison's so-called improved Phonographs, one of them being apparently a duplicate of the machine used by said Edison at his exhibition before the Electrical Club in New York in May last [see next illustration], and marked in gilt letters on the front Edison Phonograph Company, New York, U.S.A., the other one said to be the Standard [Class "M"] machine delivered to said Lippincott by said Edison…" Jesse Lippincott, president of the North American Phonograph Company, placed sample Edison Phonographs and Graphophones with both of the manufacturers in order that patented features of one machine might not be arbitrarily usurped by the other (which is exactly what subsequently occurred). In a letter of July 13, 1888, Lippincott wrote to Col. James C. Payne of American Graphophone alerting him to the imminent shipment of the Edison machine and suggesting that "…two or three of the members of your board should come over and inspect it…and they can then put a private mark on it, and later on, as soon as I can get another one, I will place the original machine in your custody." *Courtesy of the Edison National Historic Site.* (Value code: VR)

1-27. A program from the address given by Ezra Gilliland before the Electric Club, New York, on May 19, 1888. Gilliland was then at the height of his position as Phonographic innovator. Within a month, the Gilliland "Improved" Phonograph would be superceded by Edison's own "Perfected" model (which was sometimes, confusingly, called "Improved"). *Courtesy of Sam Sheena.* (Value code: VR)

Missing from this equation was any acknowledgement of the debt Edison owed to the Volta Laboratory's work. Edison simply appropriated the use of incised wax recording. Though he tried to avoid other patented Graphophone features, there was clearly no hope of achieving success in recording without incised wax. It's true that Edison's solid wax cylinder, which could be shaved and re-recorded, was an improvement of the Graphophone single-use record. Yet, the Edison system had illegally borrowed from Graphophone research.

What was shaping up was a battle of biblical proportions, which would send legions of lawyers and court personnel marching with briefs and injunctions held high, to spill their words upon the cold marble of jurisprudence.

SPITE MARRIAGE

However, there was a brief lull before the fighting started. A man named Jesse Lippincott, who had been made wealthy by his share in a glass factory, entered into an agreement with the American Graphophone Company on March 29, 1888. Lippincott, through his North American Phonograph Company, was to act as sole agent for the Graphophone. The Graphophone Company was getting ready to offer the public a talking machine for which the firm envisioned enormous demand. Although the Graphophone was intended strictly for office dictation, it was hoped that every business across the country would eventually need to lease one. Before making the machines available through North American's web of territorial, or "local," leasing outlets, the Graphophone Company planned to do one last little thing: swat that annoying Mr. Edison out of the way.

American Graphophone had Edison dead-to-rights. Edison had borrowed its method of recording without permission. Jesse Lippincott, however, stepped between the two smoldering combatants and managed to convince Edison to come onboard. Edison had recently exhibited his "Perfected" (soon to be known as the Class "M") Phonograph, the third version of the electrically driven, wax cylinder machine. The inventor was prepared to begin manufacturing these instruments, and Lippincott got the contract to distribute them. Edison, anxious to conclude a deal that would get his Phonograph back into the limelight, was forced to concede that his "Perfected" Phonograph was actually a Graphophone. The Graphophone Company then granted Edison a license that exacted a royalty on each

Phonograph delivered to North American. The two opposing factions now could embark upon their grand illusion of trying to occupy the same distributorship.

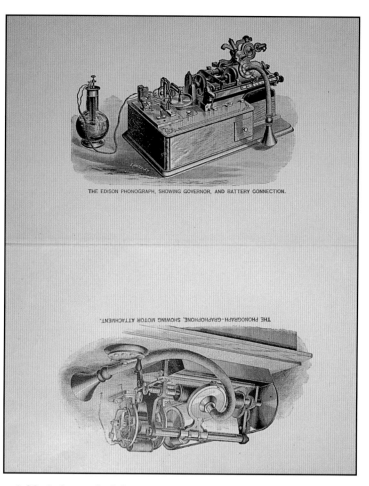

THE EDISON PHONOGRAPH, SHOWING GOVERNOR, AND BATTERY CONNECTION.

THE PHONOGRAPH-GRAPHOPHONE, SHOWING MOTOR ATTACHMENT.

1-28. Lithographed illustrations from the first months of the North American Phonograph Company (formed July 1888) showing the "Spectacle" Phonograph and the earliest commercial version of the Graphophone. The awkward terminology "Phonograph-Graphophone" at first had been considered expedient, but soon was regretted by the Graphophone interest.

1-29. An August 1888 stock certificate from the Edison Phonograph Works. Over the next 18 months, the Phonograph Works would be very busy. The North American Phonograph Company had become Edison's sole sales agency, and company documents indicate that most Edison Class "M" electric Phonograph mechanisms were manufactured prior to December 1890. *Courtesy of the collection of Howard Hazelcorn.*

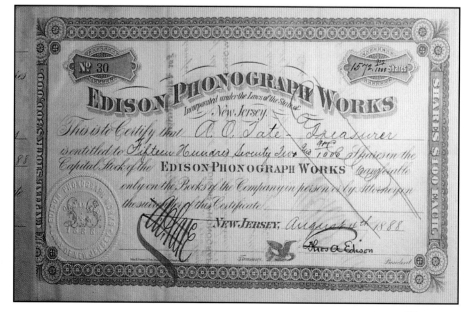

1-30. A magnificent example of the "Spectacle" style Edison Class "M" Phonograph. The "spectacle device" allowed rapid interchange of recorder and reproducer. This very early version featured sealed brass components, which on later examples were nickeled, including the pulleys, reproducer and recorder. Note also the very unusual gilt decoration similar to that on the Gilliland "Improved" Phonograph, which was not continued on subsequent Edison instruments. Early documents sometimes use the term "Improved" Phonograph to refer to this model, creating some confusion in terminology. *Courtesy of Ray Phillips.* (Value code: VR)

1-31. Colonel George Gouraud was the promoter of the Edison Phonograph in Great Britain, beginning in 1888. Edison sent Gouraud the very first model of the "Perfected" Phonograph on the day that it was completed, June 16, 1888. With this machine, Gouraud conducted musical *soirées* to which British notables and the press were invited. He sent to and received from Edison some of the earliest surviving records. The tag that accompanies these cream-colored cylinders from the Edison archives reads, "English cylinders, Col. Gouraud, Oct. 1888, keep for duplication." *Courtesy of the Edison National Historic Site.*

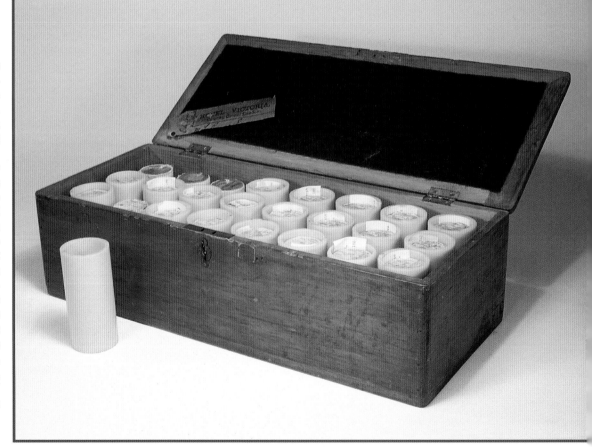

1-32. An original crate filled with Bell-Tainter cylinders. Cylinder dimensions are 1 5/16" x 6". Note the stamp on the side of the crate which reads: "New York Graphophone Co., J. K. Rho[des?] Mgr., 285 8th Avenue, N. Y. City."

1-33. Thomas Edison's employees presented him with this extraordinary machine on his birthday, February 11, 1889. The works were nickel-plated, and the "Spectacle" carriage ornately engraved, including a central "TAE" monogram. Further distinguishing this unique instrument was a gold-plated plaque, immediately in front of the motor pulley, engraved with Edison's distinctive signature. It would be nine years before Edison Phonographs would routinely carry this trademark signature. *Courtesy of the Edison National Historic Site.* (Value code: VR)

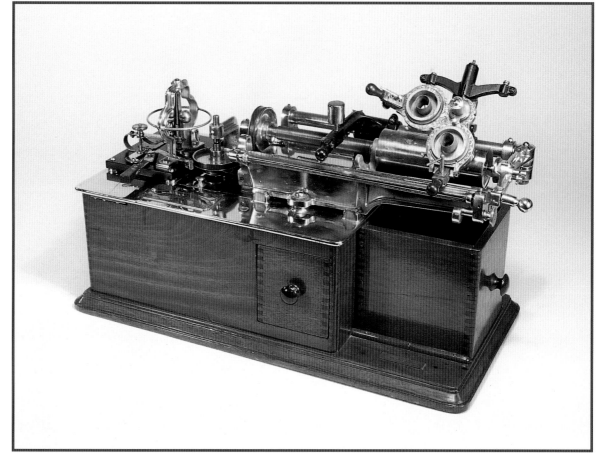

1-34. In place of a serial number, the presentation Phonograph featured the date of Edison's forty-second birthday. *Courtesy of the Edison National Historic Site.*

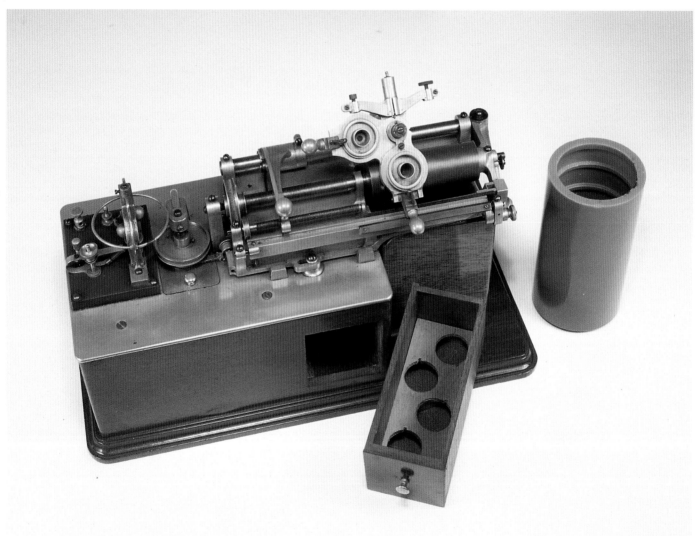

1-35. The so-called "Military" or "Portable" Phonograph of 1889 was something of a mystery. It is unclear whether the "Military" designation referred to intended use by the armed forces or by newspaper correspondents covering a war. What is evident, however, is the appealingly precise miniaturization of the mechanism. Shown next to a conventional cylinder record, the scale of this half-sized version of a Class "M" electric Phonograph becomes startlingly apparent. The drawer held four tiny cylinders, none of which are known to survive. *Courtesy of the Edison National Historic Site.* (Value code: VR)

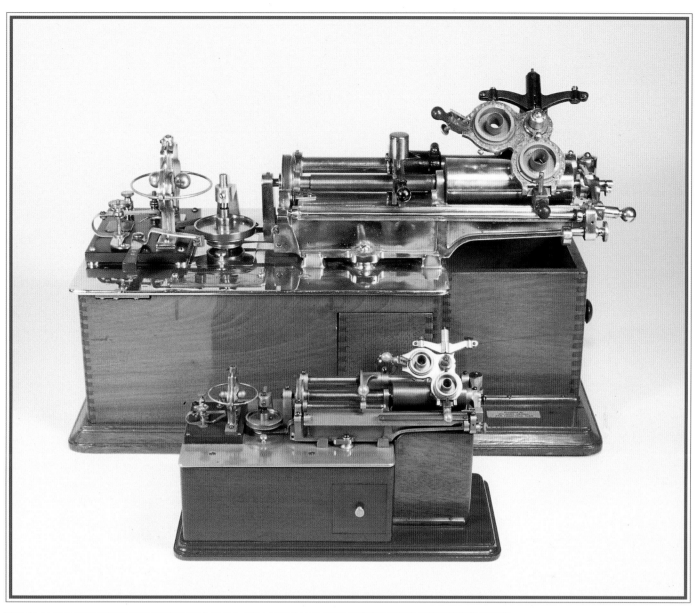

1-36. The "Military" Phonograph, shown with a full-sized Class "M" machine. *Courtesy of the Edison National Historic Site.* (Value code: VR)

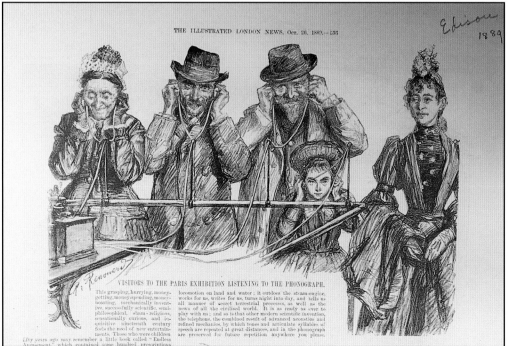

1-37. A drawing made at the Paris Exposition of 1889, which appeared in the *Illustrated London News,* October 26, 1889.

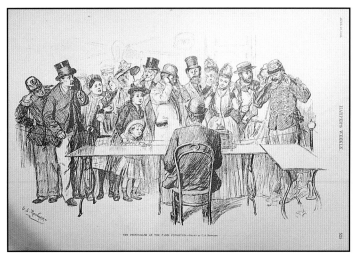

1-38. A drawing from *Harper's Weekly*, June 29, 1889, entitled *The Phonograph at the Paris Exposition.*

1-39. An early Edison Class "M" Phonograph circa 1889, retaining some features of the "Spectacle" model. The machine is housed in an unmarked oak stand of a slightly later period: possibly a custom order. The battery box may be seen on the shelf below. *Courtesy of Ray Phillips.* (Value code: VR)

1-40. The previous example with the lid in place. The serpentine drawer pulls and turned spindle legs are design elements which appeared in Hawthorne and Sheble cabinets of the mid- to late 1890s. *Courtesy of Ray Phillips.* (Value code: VR)

1-41. A close-up of this Class "M" reveals several interesting details. The straight-edge, end-gate closure and twin lugs for mounting an automatic return device are remnants of the first commercial Class "M" Phonographs. The "spectacle device" was withdrawn in November 1889, and replaced with the single carrier arm shown here. This development resulted in the revived use of "Improved" for this model. The special gilt decoration, a modified Greek-key design, has been seen on a few similar machines of this period. *Courtesy of Ray Phillips.* (Value code: VR)

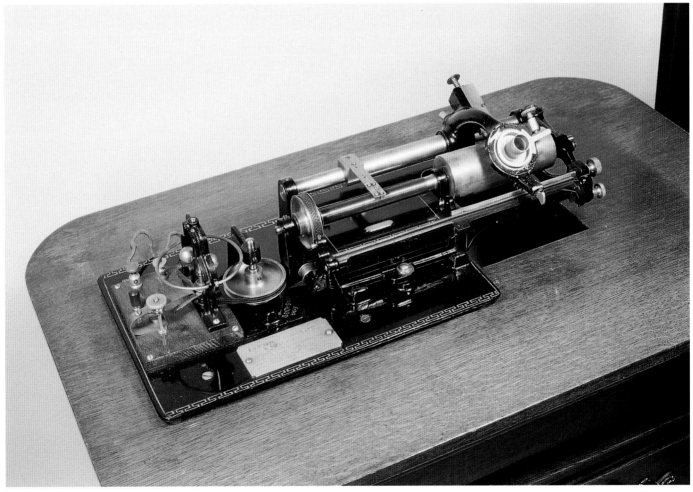

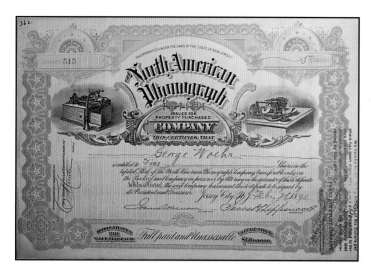

1-42. A North American Phonograph Company stock certificate signed by Jesse Lippincott. *Courtesy of the collection of Howard Hazelcorn.*

1-43. The Edison Class "T" (treadle) Phonograph was introduced in February 1889. It was an apparent answer to the treadle-driven Graphophones of the day, and the Edison version was equally impractical. Interestingly, the treadle-driven Graphophones were soon modified with battery-driven electric motors such as used by the more popular Edison Class "M," yet Edison's own treadle model survived in the catalogue until 1895. Such relative longevity was not indicative of the model's popularity: the example pictured is the only known survivor. This particular Edison Class "T" Phonograph has a fascinating history. Given by Edison himself to his sister in Milan, Ohio, it was later given to a family named Collman, from whom the present owner bought it. The example shown is in "Cabinet No. 2" (oak). A cherry version ("No. 3") was available for the same price. A September 1893 catalogue offered this model as a "Residence Outfit" with musical records for $175.00. By December, the model was sold as an "Office Outfit" with blank cylinders for $140.00. *Courtesy of Ray Phillips.* (Value code: VR)

Above: 1-44. Another rarity: Edison Class "W" (water powered) Phonograph No.5893. Introduced in February 1890, this model remained in the catalogue until 1895. However, it was not popular, due in part to a ban in some localities on excessive water use. The turbine, known as a Pelton wheel, is prominent on the left of the mechanism. To the rear of this unit, note the pipe connections and faucet. *Courtesy of Ray Phillips.* (Value code: VR)

Right: 1-45. This is the only other known complete Edison Class "W" Phonograph, and it carries unusual gilt decoration. On February 19, 1890, Edison sent a similar Water-Motor Phonograph (No.5004) to Czar Alexander III. The next day, Edison sent a Class "W" machine to a fourteen year-old piano prodigy named Josef Hofmann, "...which I have thought you would find more convenient than an electric motor machine." Two years before, Hofmann had recorded a piano cylinder at the Edison Laboratory, the first artist of note to do so. Twenty years later, he would begin a recording career with Columbia, and later, Brunswick. Along with Hofmann's Phonograph (No. 3725), Edison sent "Two dozen of musical cylinders, and fifty blank phonograms...all of which you will please accept with my compliments." It is interesting that Edison chose Class "W" Phonographs as presentation instruments, and is probably indicative of his high hopes for this design, as opposed to the messy batteries and heavy motor of the Class "M." *Courtesy of the Edison National Historic Site.* (Value code: VR)

As soon as a good number of Phonographs and Graphophones were leased to offices around the country, the swagger of the Graphophone Company began to soften. Reports from the field were not what the company wanted to hear. Edison's Phonograph was a better-designed, easier to use machine. The awkward foot-treadle of the Graphophone and its unsatisfactory single-use cylinders had quickly branded the machine a failure. The North American Phonograph Company began slowly, ever so slowly, to collapse.

The demise took five years, during which Jesse Lippincott, a man who gave new meaning to the concept of being in the wrong place at the wrong time, lost his health, his fortune and finally his life. The Graphophone Company, whose contract had been made with Lippincott personally, was frozen out of the leaderless North American Phonograph Company. The ill-conceived policy of leasing was abandoned, and North American, eventually in receivership, devolved into an exclusive sales agency for the Edison Phonograph. At the end of 1894, Edison would finally force the dissolution of North American, in order to get control of his Phonograph business. The Graphophone Company endured the long winter of its discontent, but managed to stay afloat until it could return a favor, and borrow a few of Mr. Edison's improvements. The next twenty years would be periodically punctuated by pitched battles between the Phonograph and Graphophone interests, despite cross-licensing agreements. The opening bell for what was to be a lengthy contest had rung.

1-46. The Edison Talking Doll (1890) contained the first Phonograph to employ a wax cylinder strictly for home entertainment purposes, and the first wax cylinder Phonograph to be sold rather than leased. The original base price was $10.00, a sizable sum for a child's gift at the time. Of the small number of dolls distributed, most were returned as defective, and this example is especially noteworthy for its original clothing and pasteboard carton (8" x 23"). Subsequent to the dolls' commercial failure, they were sold at a reduced price with the playing mechanisms removed and the Edison box label pasted over, as seen here. (Value code: VR)

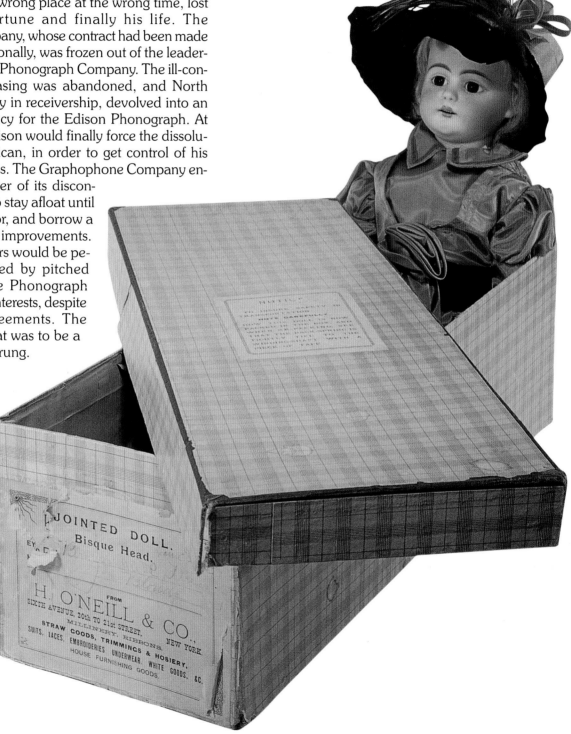

1-47. This Edison Class "M" Phonograph carries three different identification plates. To the left is the large plate found on all Class "M" Phonographs of the North American Phonograph Company period. On the Phonograph upper works is a warning that the machine was not to be used within the State of New Jersey (observing North American's sales territories). Finally, the seldom-seen plate at lower right was in response to the rampant illegal exportation of Edison Phonographs to Great Britain and Europe during the early 1890s. Legitimate distributor, the Edison Bell Company of England, requested such labeling, but North American's compliance seems to have been half-hearted at best. *Courtesy of Ray Phillips.* (Value code: VR)

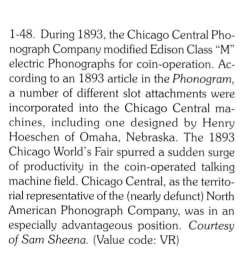

1-48. During 1893, the Chicago Central Phonograph Company modified Edison Class "M" electric Phonographs for coin-operation. According to an 1893 article in the *Phonogram*, a number of different slot attachments were incorporated into the Chicago Central machines, including one designed by Henry Hoeschen of Omaha, Nebraska. The 1893 Chicago World's Fair spurred a sudden surge of productivity in the coin-operated talking machine field. Chicago Central, as the territorial representative of the (nearly defunct) North American Phonograph Company, was in an especially advantageous position. *Courtesy of Sam Sheena.* (Value code: VR)

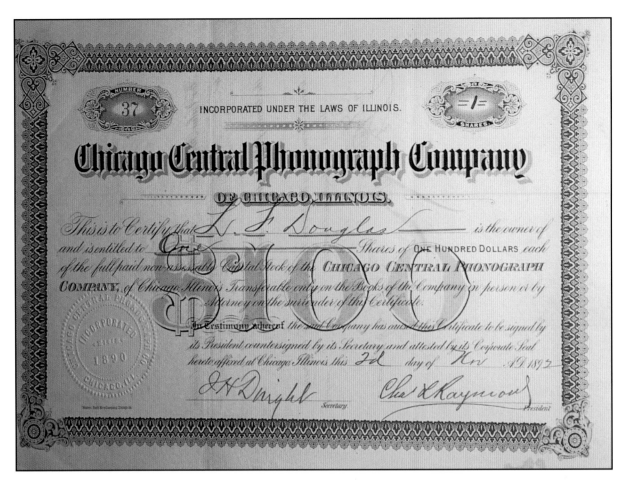

1-49. A stock certificate for the Chicago Central Phonograph Company, 7 1/2" x 11". *Courtesy of the collection of Howard Hazelcorn.*

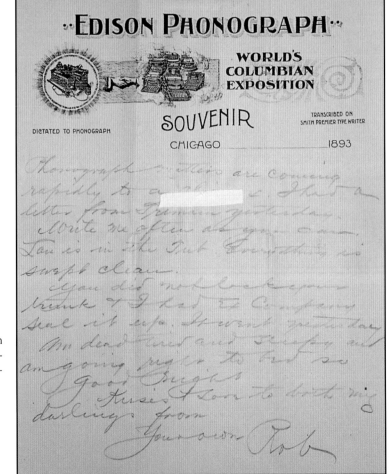

1-50. An 8 1/2" x 11" souvenir letterhead from the Edison exhibit at the World's Columbian Exposition, Chicago, 1893. *Courtesy of the collection of Howard Hazelcorn.*

EMILE BERLINER ACHIEVES THE DISC RECORD

What was Emile Berliner doing in 1887? A German mechanic and inventor who had immigrated to the United States in 1870, Emile Berliner had already designed an improved telephone transmitter, which he sold to the Bell company in 1878. The money he received allowed Berliner the financial freedom to pursue further sound experiments. However, his interest strayed from the telephone to recorded sound.

In his newly-commenced research, Berliner was initially inspired by Charles Cros. Cros' 1877 sound-recording theory had suggested converting two-dimensional depictions of sound waves into a three-dimensional (and thereby "playable") format through photoengraving. It was this theory which consumed Berliner during 1887: that a replayable record could be created out of the spiral line traced on a carbon-coated glass disc by a stylus attached to a vibrating membrane.

Although Cros, a Frenchman, was already viewed in France as the father of recorded sound, Berliner's experiments failed to vindicate Cros' photoengraving concept. Berliner could not obtain a satisfactory record by this method. It was Berliner's genius to devise a recording process, as had the Volta Laboratory Associates, employing the plastic medium of wax.

Whereas the Volta Associates had recorded in a cylinder format, Berliner, after experimenting with cylindrical recording, chose a disc. The differences between the two avenues of research went far beyond the geometric shape of the record. Volta had employed wax as the final storage medium for *vertically* recorded sound vibrations. In his 1888 experiments, Berliner used wax in quite another way. He coated a thin zinc disc with wax and set it rotating on a turntable. Sound vibrations were transferred from a diaphragm to a stylus which cut a spiral groove

through the disc's wax coating. Where the stylus removed the wax, the underlying zinc surface was exposed. Furthermore, sound vibrations were stored in the sides of the wax groove (*laterally*). By then applying acid to the disc, the lateral impressions left in the groove by the recording stylus were allowed to *etch* the exposed zinc underlay to a replayable depth. With the wax removed, this zinc plate became a "master" record. From the master, a mold or stamper could be created which could press multiple, identical copies of a performance out of a substance such as celluloid, hard rubber or (finally) shellac.

Here was the promise of Berliner's system: conveniently stackable, easily replicable sound recordings finally possible. Although at first Berliner would promote his record and the machine to play it (the Gramophone) in an extremely modest fashion, the proverbial moving finger had writ for the cylinder talking machine.

1-51. Emile Berliner developed the Gramophone or disc talking machine in the late 1880s. The very first Gramophones were manufactured as toys by the German firm of Kämmer & Reinhardt. These were distributed in Europe and Britain during the early 1890s. This example is the earliest version of the Kämmer & Reinhardt Gramophone, employing a decorated cast-iron base, self-supporting horn and hand drive. (Value code: VR)

1-52. The first disc records offered to the public were these 5" diameter Berliner "plates" manufactured in Germany for Kämmer & Reinhardt. Usually pressed of celluloid or hard rubber, these records typically carried the printed song lyrics on their reverse sides. Shown is a group of 1890-1893 K&R discs with an original shipping box.

CHAPTER TWO
1894-1903

The inventive brilliance that infused the development of the talking machine in the 1880s and early 1890s had lost its momentum. American talking machine patents had jumped to 29 granted in 1888 from only six granted in 1887. During 1889, there were 41 new sound recording-related patents, but this inventive activity had gradually dwindled to only 24 patents in 1893. A decade would pass before the 1889 high-water mark would be surpassed. The rewards for improving the talking machine had not been lucrative.

Despite significant improvements in recording and reproducing, the talking machine was viewed by the public as an expensive and cumbersome curiosity. A guidebook to the 1893 Chicago World's Fair listed 5000 exhibits and graded them "…according to their relative importance." In the Gallery of the Electricity Building, the "Latest style phonograph" (an Edison Class "M") was given the lowest grade: a "1" which denoted "Interesting."

Listed at the same location was the "First phonograph made by Edison's employes" (sic), which was graded "2," or "Very Interesting." Thus, the 16-year old "Tinfoil Phonograph" was thought, at least by the guidebook compiler, to have been more interesting than Edison's latest battery-driven Class "M" Phonograph. The North American Phonograph Company, exclusive sales agent for Edison Phonographs, no doubt would have privately concurred. Its sales of Class "M" Phonographs at $150.00 were sluggish. The machines' own shortcomings were only partially to blame. The United States was still reeling from the banking panic of 1893 and a subsequent recession. Expensive entertainment apparatus such

as Phonographs were out of reach for all but working exhibitors or the wealthy. Despite hard economic times, many Americans were willing to part with a nickel for two minutes of recorded entertainment. Had it not been for the "nickel in the slot" aspect of the talking machine business, the Phonograph might have been forgotten altogether.

Ironically, it was the relative success of the Edison Phonograph as a slot device which gave life to what became its most formidable competitor, the Columbia Phonograph Company. Originally formed as one of 32 local subsidiaries of the North American Phonograph Company, Columbia's sales territory included Washington, D.C., Maryland and Delaware. Columbia soon discovered, as did the other local companies, that the Edison Phonograph was difficult to lease or sell, and the Bell-Tainter Graphophone was virtually impossible to market. Far more lucrative were the recording, duplication and sale of musical cylinders.

Columbia was undoubtedly the most aggressive merchandiser of pre-recorded musical cylinders throughout the 1890s. Remarkably, this preeminence was attained while most of North American's other subsidiaries were floundering. Columbia's success was due in large measure to the foresight and sagacity of its president, Edward D. Easton, who quickly recognized the importance of supplying "non-durable" goods to the market. Most of Columbia's records during this period serviced coin-operated Edison Phonographs. This early symbiotic relationship would, by the late 1890s, degenerate into bitter rivalry.

2-1. The Graphophone, developed by Edison's rivals under the leadership of Alexander Graham Bell during the 1880s, proved impractical to operate until it was adapted in the early 1890s to improve its performance. The original format of Graphophone cylinders (known as Bell-Tainter records) was a paper tube 1 5/16" x 6", thinly coated with wax. These were abandoned and the Edison format adopted: a solid wax cylinder of 2 1/8" x 4". At the same time, the foot treadle power of the first Graphophones was replaced by a small electric motor, allowing the Graphophone to begin a slow ascent toward serious competition with Edison. The Type "R" Graphophone shown here (1894) is an especially rare model which featured a repeating attachment. (Value code: VR)

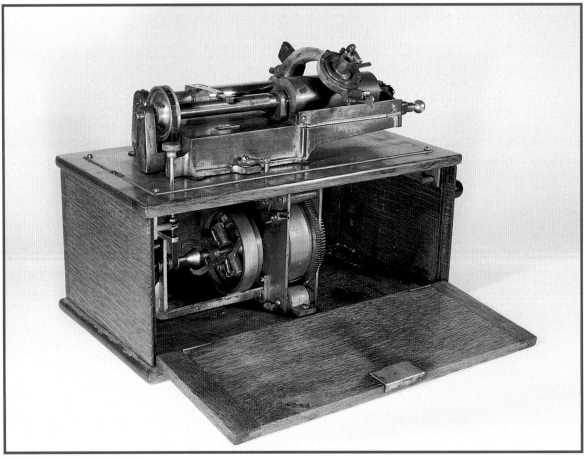

2-2. In 1894, Edward H. Amet introduced a spring motor to power both Phonograph and Graphophone mechanisms. The very first models used a single mainspring. Pictured is the earliest type of Amet motor, coupled with an Edison Class "M" Phonograph upper works (with optional nickel-plating). The governor of this particular motor, which was also employed in certain double-spring motors to follow, much resembled a flywheel, but was in fact a genuine friction governor with centrifugally-activated weights. (Value code: VR)

2-3. When the early Graphophones were modified to accept the Edison-style cylinders, a supplemental mandrel was necessary to support the records. However, the American Graphophone Company also offered a Type "E" cylinder (shown) from December 1892 until September 1897, which consisted of a Bell-Tainter paper core sufficiently coated with wax to conform to the approximate outside diameter of an Edison record. These could only be used on the earlier "Bell-Tainter" style Graphophones, and when newer models were introduced, these cylinders became obsolete. Although commercially recorded examples were never offered, these 6" Type "E" cylinders were capable of holding over seven minutes of recording at 100 rpm, far exceeding the 4" Edison cylinders they imitated. *Courtesy of Lawrence A. Schlick.*

2-4. The Edison "Kinetoscope" was the world's first moving picture viewer. These "peep-show" arcade machines, which showed a loop of film when activated by a coin, were introduced in 1894. Thomas Edison, during the development of his motion picture system, had naturally envisioned linking sound with image. Although the road to successful synchronized sound was to prove long and laden with obstacles, some "Kinetoscopes" were fitted with specially modified Edison Phonograph mechanisms. These machines, called "Phono-Kinetoscopes," included background music for the visual action. An incomplete Phonograph works from a "Phono-Kinetoscope" is shown here. At the right front, note the unusually short width of the record repeater's coarse feed screw. Since a "Kinetoscope" film loop (approximately 50 feet in length) lasted for 60 seconds or less, and a conventional cylinder record played for about two minutes, only a portion of the record was necessary to contain the accompanying music. Existing records with appropriate musical content might simply have been excerpted by the mechanism, or specially prepared cylinders of shorter playing time might have been distributed with the machines. The same electric motor which drove the film ran the Phonograph and repeater. *Courtesy of Ray Phillips.* (Value code: VR)

2-5. On November 30, 1892, George Tewksbury filed a patent on a "Coin-Operated Mechanism for Phonographs." The device was reasonably successful, and known in the trade as the "Kansas" mechanism, due to its exploitation by Tewksbury's Kansas Phonograph Company. Once the patent was granted on July 24, 1894, a small brass plate was affixed to the bedplates of Edison Class "M" Phonographs equipped with the device. This "Edison Automatic Phonograph" of the 1894 period features the "Kansas" mechanism, which was later marketed by the United States Phonograph Company of Newark, New Jersey. This example also displays a rare original sign insert. *Courtesy of the Sanfilippo collection.* (Value code: VR)

THE GENIUS BEHIND COLUMBIA'S SUCCESS

One other man, besides Easton, loomed large in the fortunes of the Columbia Phonograph Company and the American Graphophone Company, which were consolidated in 1895. Thomas Hood Macdonald entered the talking machine industry in 1889 when he was hired by the Eastern Pennsylvania Phonograph Company in Philadelphia, one of North American's local subsidiaries. Macdonald's mechanical talent was quickly recognized, and he was made "General Inspector" of the North American Phonograph Company in August 1889. From December 1889 until July 1890, Macdonald was stationed at the factory of the American Graphophone Company in Bridgeport, Connecticut, to receive Graphophones manufactured for North American. During this period, the vice-president of North American, Thomas R. Lombard, asked Macdonald to duplicate the secret Edison wax cylinder formula. The factory manager of the American Graphophone Company, N.E. Russell, offered the use of the company's Bridgeport facilities for the project. Macdonald agreed, set to work, and by April 1891, had become the "chief experimentalist" for the American Graphophone Company.

The following year, Macdonald was made acting manager of the American Graphophone Company factory, and was promoted to manager in 1893. The affairs of the American Graphophone Company during this period were conducted on a shoestring budget. Sales of Graphophones had nearly ceased. Edison was the major supplier of blank cylinders to the trade, and companies such as Columbia made most of their recordings on Edison blanks. American Graphophone centered much hope upon Macdonald's effort to develop a cylinder blank of comparable quality to the Edison product, and which would provide it with much-needed cash.

By December 1892, Macdonald had developed a "lead soap" which was used to make blank cylinders. This formula was first used by American Graphophone in the 6" long x 2" diameter Type "E" cylinders, which incorporated the old Bell-Tainter paper core. Soon afterward, American Graphophone began the manufacture of standard-sized cylinder blanks, which were sold to the Columbia Phonograph Company. Problems with "clouding" or oxidation were brought to Macdonald's attention, and his contract to provide blanks to American Graphophone (at 1 cent each) was in jeopardy. Seeking expert help, Macdonald placed an advertisement in the September 1894 issue of the *American Soap Journal:*

Wanted: Thoroughly practical man, capable of carrying on experimental work in hard soap making. Work is on a metallic insoluble soap, not used for washing purposes. One versed in the working of stearine, waxes and lead soaps, greatly to be preferred.

Address, T. H. Macdonald, Manager, Bridgeport, Conn.

On September 14, 1894, Adolph Melzer, the proprietor of a soap factory in Evansville, Indiana, answered Macdonald's query, offering to undertake the experimentation at no cost if he (with brother Charles) failed, but expecting remuneration corresponding to the value of their time and labor if successful. Macdonald, operating under a January 1, 1895 deadline from American Graphophone to "…furnish a suitable wax…or have the contract cancelled and pay the damage resulting from the loss of music and other records placed on the soft cylinder," immediately accepted Melzer's offer. The two men began a voluminous correspondence focused on chemical formulae and cylinder blank molding techniques.

Numerous samples of wax composition and cylinders were exchanged between Bridgeport and Evansville. As Macdonald's deadline loomed ever closer, he pressed Melzer for completion of the formula and an accounting of expenses. Melzer, unaware of Macdonald's circumstances, laconically replied that he and his brother wished to postpone an accounting until they could "…get it entirely right…" and so that Macdonald would not have to "…put up with a little unnecessary brittleness &c &c." Macdonald, however, insisted on an accounting, and on December 8, 1894, Melzer replied:

Dear Sir: Your telegram of 8th inst. is just to hand. Our time, labor and materials that we have employed and results obtained in experiments we made for you, are worth at a low estimate $500.

Macdonald responded in a rather alarmed tone:

It strikes me that the figure you have named, $500 is pretty high. Can't you review the matter and give us a little better figure? We find it pretty hard to struggle up with a new business on our hands, and $500 in one lump is pretty large to us.

This statement is indicative of American Graphophone's financial affairs at the time. Melzer, believing that he had been dealing with an established and prosperous organization, replied on December 18, 1894 with some condescension:

Rather than accept one cent less than the amount named, we should prefer to take no money at all. Send us a nice graphophone for our parlor and we will tell you how to make the composition for your cylinders.

Macdonald, somewhat embarrassed, responded on December 22, 1894, expressing his appreciation for the work performed by Melzer and the "first class" results he had achieved. At the same time, Macdonald took the opportunity of the bargain offered by Melzer:

I most gladly accept your offer to receive one of our machines in payment. I am sending you today one of our most complete outfits….I send full directions with the machine and have tagged the different parts so that there will be no difficulty in using same. We have none of the parlor cabinets now in stock, so I send the machine in a small case. We will send you one of them as soon as they are finished and you can place the machine in it.

Graphophones of this period were offered for approximately $130.00, while the parlor cabinet was priced at $35.00. Customers were rare. For less than $200 of virtually unsalable goods, Macdonald had settled a $500.00 debt. Melzer's Graphophone arrived on Saturday, December 29, 1894. Two days later, a delighted Adolph Melzer wrote to Macdonald:

Everything opened up all right and we kept that "Phone" humming the greatest part of the day yesterday. Please accept our best thanks; we are almost glad now you did not send us a check, for we can now understand your needs a great deal better, and whilst we will make no promises, think we can further improvements in the cylinder composition and will continue our experiments.

There followed additional numerous and rapid exchanges of letters full of detailed chemical formulae. Melzer and Macdonald would experiment following the other's suggestions and shoot back a report of the results, including suggested steps for the other to take. This soon led the two men into areas outside the realm of chemical compounds. Manufacturing techniques, cleanliness, filtering, and the differing qualities of available raw materials were discussed in minute detail. As more progress was made, relations between Macdonald and Melzer warmed. Macdonald visited Melzer's Evansville soap factory in early February 1895, absorbing more information on the manufacture and molding of soaps.

On March 29, 1895, Macdonald wrote to Melzer:

At last I have succeeded in getting your machine with parlor cabinet off to you. I shipped by freight via the Pennsylvania. The machine is in the box in the battery compartment of the cabinet. Take out the screws in the bottom and back which hold the box in place. Then lower the front door of the compartment and slide the box out. The machine is to be set in the hole in the top of the cabinet so that the plugs in corners of the plate will rest in the rubber buffers. You will note that the top of the cabinet lifts off when unlocked. The extra parts are in the blind drawer, and can be taken out through the machine hole.

This was the second Graphophone sent to Melzer

by Macdonald. There is no evidence of a request by Melzer for an additional machine, but his ongoing work for Macdonald no doubt warranted subsequent remuneration. Macdonald, for his part, was undoubtedly pleased to pay with such relatively worthless specie for such valuable help.

During the early months of 1895, Macdonald's letters became more candid about duplicating the Edison cylinder formula, which he obviously considered superior to his own. Macdonald relentlessly queried Melzer about certain ingredients purportedly used in the Edison recipe, particularly acetate of alumina. Melzer patiently replied to each inquiry with the admonition to avoid acetate of alumina, and often accompanied by cylinder blanks made with the ingredient to demonstrate its flaws. Melzer recommended hydrate of alumina and soda in solution, and sent hundreds of samples of various mixtures for Macdonald's approval. A box of cylinders was sent to Macdonald on March 31, 1895, with a letter stating:

We consider these cylinders best up to date, and if a little more of the hydrocarbons would not make them too gummy, they would be less noisy. You will see we have made these cylinders 5 ¼" long…

These Melzer cylinders, made only for his own use, represent an early attempt to increase the playing time of Edison-type cylinders. The 5 ¼" cylinders physically extended beyond both ends of the Edison-style mandrel supplied with the Graphophone. These early Graphophones were originally designed to play cylinders of 6" length, allowing Melzer's cylinders to be easily accommodated. The design of Edison Phonographs prohibited the use of cylinders exceeding 4" in length. In the same letter, Melzer suggested a proposition:

The principal of our High School, Mr. Spear…is after us to give an exhibition at the school; we cannot refuse and have agreed to give them an exhibition next Thursday week (Apr. 11)….We believe it would be a capital "hit" for you to deliver the lecture part on one or more of the large [5 ¼" length] cylinders we sent you, and thus let you do the talking to those luminaries for us. Make the lecture as long, strong, and elaborate as you can. In addition to this, we would like to have about a dozen good records of classical music on the large cylinders, taken at a speed of 115-120, and if you will attend to this for us, so that we will have the cylinders here in ample time, we will be greatly obliged to you.

2-6. In September 1894, the factory manager of the American Graphophone Company, Thomas Macdonald, advertised for a "Thoroughly practical man, capable of carrying on experimental work in hard soap making…" Macdonald, attempting to develop an improved wax composition similar to the Edison formula, was a desperate man. For over two years he had worked on his wax formula, only to have the finished cylinder blanks continue to "cloud" after a few weeks. American Graphophone, hoping to generate badly-needed income through the sale of its own cylinder blanks, had given Macdonald an ultimatum: furnish a suitable wax by January 1, 1895, or "…have [the cylinder] contract cancelled and pay the damage resulting from the loss of music and other records placed on the soft cylinder." Adolph Melzer, the owner of a soap factory in Evansville, Indiana, responded to Macdonald's plea for help. Under Melzer's tutelage, Macdonald's wax cylinders were improving from week to week. As payment, Macdonald sent Melzer a

Graphophone on December 22, 1894, and another on or about March 29, 1895. By that time, Melzer was molding cylinders for his own personal use, 5 1/4" in length. The machine pictured is one of Melzer's Graphophones, modified with an Edison "Triton" spring motor in a purpose-made cabinet. Melzer evidently enjoyed giving local concerts with his own recordings of local talent. Shown are some of Melzer's 5 1/4" long cylinders along with those of standard 4" length. Note that the "Triton" spring motor was situated so that it could be wound from behind the machine as it was being exhibited. *Courtesy of the Charles Hummel collections.* (Value code: VR)

2-7. A program from one of Adolph Melzer's "Graphophone Concerts" dating from June 3, 1895. Mention was made of the local talent employed by Melzer, as well as the large size of his cylinder records. The Evansville businessman apparently enjoyed his leisure pursuit for several years, but it was a small price for American Graphophone to pay in order to become the premier cylinder record manufacturer of the 1890s. *Courtesy of the Charles Hummel collections.*

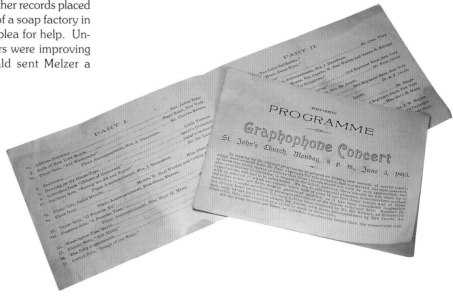

Macdonald, ever anxious to please his benefactor, made the requested recordings and dispatched them to Evansville. At the same time, he persisted in his inquiry surrounding the form of alumina that the "Silver Lake People" (Edison's manufacturing staff of Silver Lake, New Jersey, which was outside of West Orange) were using. Pages of manuscript explored various possibilities to the answer of what had clearly become an obsession with Macdonald: discovering the chemical composition of Edison cylinders.

Melzer's letters to Macdonald were models of patience and instruction, yet after a particularly long and detailed discussion of the properties of the various forms of alumina, he burst out: "We wish that we could make this so clear to you that you could not help but see it and *believe* it." Always encouraging, Melzer nevertheless put his finger on the problem:

In conclusion, don't get desperate, and waste no time or money trying to find out how the S. L. people do it; your troubles are purely of a mechanical, not chemical nature...

On April 14, 1895, in the face of Macdonald's unceasing references to the "S.L. cylinder" and the use of acetate of alumina, Melzer wrote:

The trouble in your case, it is clear, is that you have too many things to attend to; the S. L. people make the comp.[osition] only; don't even mold it; whilst you are attempting to do both with imperfect and inadequate apparatus, superintend the mechanical part of your factory, and study out new types of machine besides. That is more than any one man can do well. The way we have taught you to make the comp. is right, and to use the acetate or any other salt, and dry soda, is wrong.

Macdonald remained unconvinced for quite some time, but after numerous experiments and exchanges of letters, he finally admitted to Melzer:

...I have come to the conclusion that the so-called dry mixture, that is made with acetate of alumina, is of no use. It is beginning to sweat. This is enough. We don't want to touch it with a 40 foot pole. It has acted just as you said it would.

The presence of "pinholes" in the cylinder surface remained a problem, and Macdonald appealed to Melzer for assistance. Melzer reiterated his belief that the pinhole problem was based in faulty molding technique. He offered to enter into an arrangement with American Graphophone to manufacture cylinder blanks, or, if this was not suitable, to visit Bridgeport, "...and spend a week or two with [Macdonald] to relieve him of his trouble with the 'pinholes'." Macdonald enthusiastically accepted Melzer's offer to visit the American Graphophone factory. Melzer arrived at Bridgeport on July 19, 1895, and worked with Macdonald every day until the 30th, by

which time the "pinhole" problem had been rectified. Melzer, as he had predicted, found numerous faults with Macdonald's Bridgeport cylinder plant, which he later described:

The "pin holes" complained of by Mr. Macdonald were evidently caused by the lack of care in the manipulations of molding these cylinders. I found the conditions under which Mr. Macdonald worked rather primitive. The vessels used for pouring liquid composition were open tin vessels which discharged their contents from the top, thus permitting all air bubbles of which there are generally more or less on the surface, to run into the molds, there to be arrested in the congealing mass to form flaws, "pin holes." In place of the tin vessels or dippers discharging from the top I ordered and used tin vessels having a spout starting from near the bottom of the vessel, same as in the case of the ordinary tea-pot...

Melzer further noted that Macdonald had no provision for quickly cooling the molded cylinders to room temperature. Melzer's attentions transformed Macdonald's cylinder molding plant into a more viable enterprise. So successful, in fact, that Macdonald was prompted to write Melzer on August 11, 1895: "Cylinders seem to be going along all right. Pinholes are very scarce, so much so as to be of no consequence whatever."

By this time, the American Graphophone Company and the Columbia Phonograph Company had consolidated their management, and Columbia cylinders rapidly assumed trade preeminence which would endure for several years. Thomas Macdonald, whose mechanical genius was made tangible in several classic talking machine designs, owed much of his cylinder record manufacturing knowledge to Adolph Melzer, to whom he had written:

I hardly know what to say to fully express my appreciation of your most excellent work in our behalf....In my opinion your mixtures, with the possible exception of the first samples, are fully the equal of anything ever turned out by the [Edison] phonograph Co. and *several samples I consider infinitely superior to them.* There is to me a *virility,* a *brilliancy* in the record that those on [Edison] phono. cylinders do not possess.

Macdonald filed a patent on "his" cylinder molding technique on November 27, 1896. While testifying in a related case in 1906, Melzer was asked: "Did you ever remonstrate with Mr. Macdonald orally or in writing for having taken the patent in suit without your knowledge; and if not, why not?" Melzer replied:

I did not; I was willing that he should have all the benefits from that invention, but felt that he did wrong in taking out that patent and bodily claiming our (my brother's and my) formulas and process for the composition as his own without consulting us and obtaining our consent.

Macdonald himself admitted that of all the "various chemists and experimenters" he had consulted in regard to cylinder composition and molding techniques, "I did not consider the suggestions of any of the people I have mentioned of any value except those of Mr. Adolph Melzer."

2-8. Louis Glass of San Francisco had been the manager of one of the local subsidiaries of the North American Phonograph Company. In addition to patenting the first coin-operated Phonograph mechanism in 1889, Glass designed a spring motor for Phonographs offered in 1895. Shown is a Phonograph (No.3664) equipped with this motor, with the winding crank protruding from the front of the cabinet. A mainspring tension indicator may be seen above the crank. *Courtesy of Ray Phillips.* (Value code: VR)

2-9. A close-up of the Glass spring motor, showing the horizontally-mounted mainspring. The motor is marked: "Edison Phonograph Co., 430 Pine St., S.F." An example exists with the governor mounted beneath the bedplate. *Courtesy of Ray Phillips.* (Value code: VR)

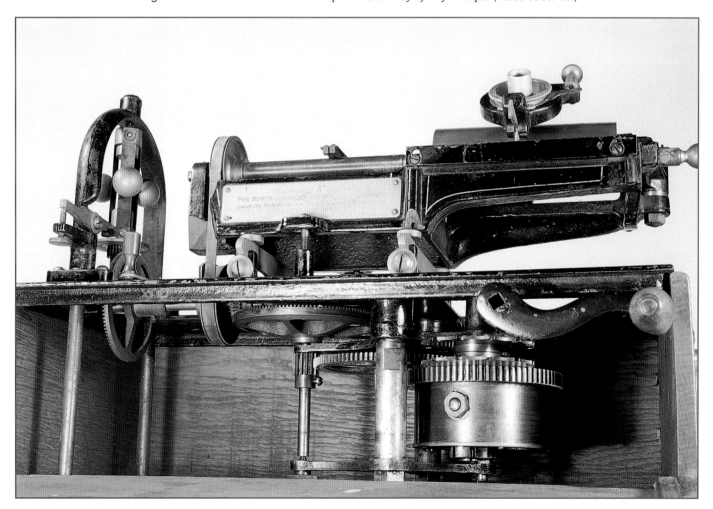

2-10. When Emile Berliner was attempting to find investors for his Gramophone in 1894 and 1895, a recurrent problem was the widely held belief that his device was simply a "toy." The earlier European incarnation of the Kämmer & Reinhardt Gramophone as a plaything continued to haunt Berliner's efforts to legitimatize his disc talking machine. One way to combat this prejudice was by providing the Gramophone with an electric motor, as was used in the Edison Class "M" Phonographs of the day. A few documents of the 1894 period depict Berliner Gramophones with electric motors, but not until the development of a reliable spring motor would the device receive widespread acceptance. The instrument shown is a later (1896-97) Berliner hand-driven model, modified with an electric motor of the period. *Courtesy of the Domenic DiBernardo collection.* (Value code: VR)

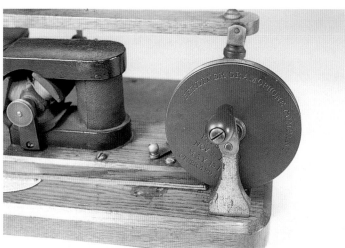

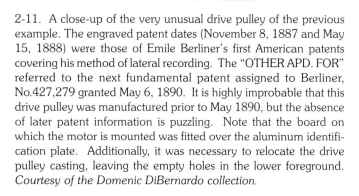

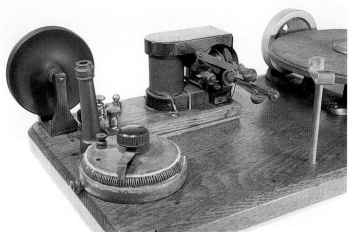

2-11. A close-up of the very unusual drive pulley of the previous example. The engraved patent dates (November 8, 1887 and May 15, 1888) were those of Emile Berliner's first American patents covering his method of lateral recording. The "OTHER APD. FOR" referred to the next fundamental patent assigned to Berliner, No.427,279 granted May 6, 1890. It is highly improbable that this drive pulley was manufactured prior to May 1890, but the absence of later patent information is puzzling. Note that the board on which the motor is mounted was fitted over the aluminum identification plate. Additionally, it was necessary to relocate the drive pulley casting, leaving the empty holes in the lower foreground. *Courtesy of the Domenic DiBernardo collection.*

2-12. Another view showing the electric motor governor and rheostat. The Berliner stamp is partially visible at left, citing a February 19, 1895 patent date. This suggests the earliest possible period for this particular machine. The anodized castings are believed to date from 1896-97. Contemporary illustrations (circa 1894) of electric-motor Berliners showed a direct drive from the motor armature via a small wheel to the underside of the turntable. Additionally, the documented electric motor employed a vertical governor. However, the May 16, 1896 issue of *Scientific American* pictured a horizontal governor, string-drive configuration much like that shown here. This example may be an experimental model, or an early modification made by the owner to improve its performance. In either event, this is one of the earliest electric motor Gramophones in the world. *Courtesy of the Domenic DiBernardo collection.*

The Cylinder Talking Machine Business

By late 1895, the Columbia/American Graphophone alliance had taken a clear lead in the cylinder talking machine business. The popular Type "N" Graphophone was introduced in September of that year for $40.00. In March 1896, Edison began offering his "Spring-Motor" Phonograph, using the Capps "Triton" motor, for the sobering price of $100.00. Several weeks later, Edison introduced the "Home" Phonograph for $40.00. However, the first "Homes," equipped with a small motor supplied by an outside manufacturer, were under-powered and poorly regulated. Sales of Edison products languished while Columbia continued to prosper.

The $25.00 Columbia Type "A" was introduced in December 1896. Edison could offer no answer other than an improved "Home" with a more powerful motor for $40.00. By September 1897, Columbia had introduced the three-spring Type "C" Graphophone for $50.00, and the tremendously successful Type "B" Graphophone (called the "Eagle") for $10.00. The initiative shown by Columbia in successfully lowering the cost of talking machines left Edison far behind, despite an August 1897 reduction in the price of the "Home" to $30.00.

2-15. A close inspection of this Phonograph upper works mounted on a Graphophone ("Macdonald" design) motor reveals that the upper casting is aluminum, and obviously a direct copy of an Edison. The manufacturer of this upper mechanism is not known, but the use of aluminum suggests that it may have been a product of the Chicago Talking Machine Company, whose innovative aluminum motor appeared in 1896. (Value code: VR)

2-13. The Type "N" ("Bijou") Graphophone was introduced in 1895 at $40.00 and immediately had a substantial impact on the talking machine market. Columbia was bent on putting the talking machine into the hands of the public, and the "N" represented the first "modestly" priced machine ever to be offered. The sturdy spring motor which ran the "N" made the older battery motors unnecessary, and very few "N"s were fitted for electrical operation. Shown here is Type "N" serial No.40,064, from about the first month of production, September 1895 (numbering began in the 40,000 block). It was equipped with the same type of two volt motor which had powered Graphophones since 1893. The cabinet was finished without a decal, as occasionally was seen in a previous model, the Type "G." When Type "N" production commenced in full, a Graphophone banner decal would become customary. (Value code: E)

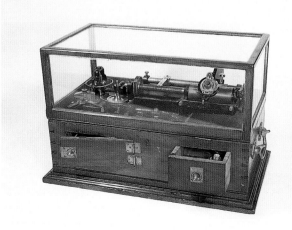

2-14. At first glance, this appears to be an Edison Class "M" Phonograph in a British custom glazed cabinet, circa 1895. However, closer examination reveals that the machine is an extremely fine copy, locally produced, right down to the reproducer and recorder. Craftsman-like imitations of Phonographs and Graphophones were rather common in the 1890s, when talking machines were expensive and mechanical labor was not. Most, though, were executed in plain brass. The fact that black-enameled castings have been used here is quite extraordinary. *Courtesy of Sam Sheena.* (Value code: B)

2-16. This intriguing coin-operated Edison Class "M" Phonograph was listed in a Hawthorne and Sheble catalogue of 1896 as the "improved Nickel-in-Slot Phonograph." "Any one who places one or more of these outfits in a store, hotel parlor or corridor, billiard hall, saloon, depot, or any place where people pass or congregate will find they… will earn handsome profits right along." At least two slight variations of the machine were offered: one which required the customer to activate a push-button after he dropped the coin, and one which was automatically started by the coin. Outfit No.12 (the push-button style) included the machine, electric battery to run it, twelve cylinder records, and various accessories for $200.00. The automatic-start outfit sold for $225.00, but three years later was being offered without records at only $125.00 by dealer T. J. Moncks of New York City. *Courtesy of Sam Sheena.* (Value code: A)

2-17. In the spring of 1896, a new $40.00 Edison Phonograph was announced. Known as the Edison "Home" Phonograph, it was initially under-powered and inadequately regulated. These earliest models were powered by a small clockwork motor that employed a string to drive the large upper pulley. All indications suggest that the first "Homes" were an unmitigated failure in the marketplace. This example, the earliest known, is No.138. *Courtesy of Ray Phillips.* (Value code: VR)

2-18. The tiny motor of Edison "Home" Phonograph No.138. Note the two small cylindrical governor weights located between the two mainsprings, and the combination on/off/speed control. *Courtesy of Ray Phillips.*

2-19. Another view of the earliest Edison "Home" Phonograph motor. The two mainspring barrels may be seen, as well as the string drive. *Courtesy of Ray Phillips.*

2-20. The "Multiplex," when first introduced in 1896, was an attachment designed to substitute for the Edison Phonograph upper works most commonly found on Class "M" and "Spring-Motor" Phonographs of the time. An article in the November 1896 *Phonoscope* described the "Multiplex" as follows: "This attachment consists of a frame containing five mandrels which can be attached to the standard [Class "M"] Edison machine. A record can be put on each mandrel and any one reproduced at will by simply bringing it into place, which is done by means of a compound reacting ratchet lever, the knob of which is exposed in the front part of the machine. It is especially designed to slot machines, as anyone patronizing such a Phone with this attachment on can have their choice of five selections. It is also very valuable for commercial purposes, as five cylinders may be dictated to at one sitting, without changing cylinders." The advantages of the "Multiplex" as applied to coin-operation are readily apparent, but less obvious are the benefits as embodied in a machine for business dictation, as shown here. A self-contained unit complete with spring motor, this machine was designed to avoid the exasperation described in the December 1896 *Phonoscope*: "The whole trouble was in the single cylinder, which did not allow sufficient space for the mass of dictation…associated with the routine of daily business. When the space on one record blank was reached, the only alternative was to remove the cylinder and insert a new blank. This occasioned no little annoyance, and consumed valuable time…" This particular example was equipped with six rubberized mandrels. The 1358 Broadway, New York, address on the front of the machine was that of George V. Gress (and sons). Gress was an Atlanta, Georgia, entrepreneur who became partner to George W. Moore, also of Atlanta. Moore was granted a patent for the "Multiplex" on September 22, 1896. The two had a falling out, and the partnership was dissolved in the spring of 1897. The April 1898 issue of *The Phonoscope* announced the Multiplex Company's move to 1358 Broadway. One year later, the company moved again, which suggests the approximate period for this particular machine. (Value code: VR)

Edison countered in March 1898 with a brilliant design called the Edison "Standard" Phonograph, introduced for $20.00. A substantial machine, the "Standard" would prove to be the most enduring of all Edison Phonograph designs. Columbia, not to be outdone, introduced two additional popular models to its line: the $25.00 Type "AT" and the Type "Q," priced at an astoundingly low $5.00.

The cylinder record was the predominant talking machine format, but the introduction of the Graphophone Grand (Type "GG") in December 1898, marked the first chink in the cylinder's armor. Despite the Type "GG's" impressive price (introduced at $300.00, soon reduced to $150.00) and the $5.00 initial price of its 5" diameter cylinder records, this development served to clearly illustrate the drawbacks of the cylinder format. The "Grand" cylinders were fragile, cumbersome, and occupied a lot of space. These disadvantages were present to a lesser extent in the smaller cylinders, but in comparison to Berliner's flat, durable 7" disc records, the cylinder format began to show its flaws. A stack of sixty Berliner discs occupied roughly the same space as four standard-sized cylinders. In February 1899, Edison announced his version of the unwieldy 5" cylinder machine, the "Concert" ($125.00), as well as the budget-priced "Gem" ($7.50). It was the golden era of the cylinder, but the end was already in sight.

THE DISC TALKING MACHINE BUSINESS

The disc talking machine market during the 1890s was the almost exclusive domain of Emile Berliner's Gramophone. In 1894 the Gramophone business had tenuously begun with the opening of a small store in Baltimore, Maryland, offering only crude hand-driven machines and 7" discs recorded by a primitive acid-etching process. The Berliner Gramophone Company was established in October 1895, but with painfully limited sales to sustain it. In 1896, advertising man Frank Seaman contracted with Berliner to be exclusive sales agent in the United States. Seaman's National Gramophone Company began running advertisements in popular magazines, and sales of the $15.00 hand-driven Gramophones began to pick up. More importantly, a lever-wound spring-driven version of the Gramophone, developed by Levi H. Montross and mechanic Eldridge R. Johnson, was introduced for Christmas 1896 at $25.00.

In early 1897, the cabinet of the spring-driven Gramophone (still lever-wound) was redesigned and given an improved soundbox devised by Eldridge Johnson and Alfred Clark. By August 1897, Johnson had redesigned the motor, governor and brake, and applied for a patent on what was to become known as the "Improved" Gramophone. This machine, selling for

$25.00, would single-handedly enable the Gramophone to become a viable competitor of cylinder talking machines. In addition, the "Improved" Gramophone, along

2-21. The problem of wavering pitch had been associated with hand-driven talking machines since the days of "Tinfoil Phonographs." Although Berliner continued to offer hand-driven Gramophones alongside spring-driven models, the hand-driven mechanisms did little to enhance the prestige of the company. An attempt to absorb erratic hand rhythm or "jerkiness" is seen in an improvement to the model shown: a crude slip-clutch incorporated into the flywheel. Alfred C. Clark, co-inventor of the soundbox used on "Improved" Gramophones, filed the patent application for this device on December 4, 1896, which was granted (No.597,875) on January 25, 1898. (Value code: VR)

2-22. A close-up of the Berliner slip-clutch. Power was transferred from the nickeled pulley to the flywheel through contact with a leather shoe at the upper right. Friction between the pulley and flywheel was adjustable by loosening the screw at lower left and rotating the arm that was attached to the friction shoe by a coil spring. Once the turntable was brought to operating speed, the flywheel/slip-clutch assembly would tend to compensate for uneven hand movement.

with a dog named Nipper, would eventually become one of the world's best-known trademarks: "His Master's Voice."

By late 1897, Frank Seaman had come to the same realization as have countless modern-day collectors: that the terms "Gramophone" and "Graphophone" were liable to be confused. Seaman began marketing the Berliner "Improved" Gramophone as the "Zonophone" or the "Vocophone." Berliner took a dim view of such monikers, and Seaman was instructed to immediately resume use of the term "Gramophone." This would not be the last word on the matter.

2-23. A Berliner record catalogue from the 1897 period. *Courtesy of the John and Nancy Duffy Family.*

The fall of the Berliner Gramophone Company in the United States was precipitated by the arrogance and limited vision of its own management. Frank Seaman made several attempts to arrange for the manufacture of less expensive, more efficiently designed Gramophones, as allowed by his 1896 contract. The Berliner management flatly refused to consider any manufacturer other than Eldridge Johnson. Seaman quite rightly surmised that covert financial ties had developed between the Berliner management and Johnson. Judging by Seaman's persistence, he evidently believed that a superior sample machine at less cost than Johnson's "Improved" Gramophone would eventually win over the Berliner interest.

In February 1898, Seaman launched the Universal Talking Machine Company. Its talented inventor, Louis Valiquet, designed a high-quality disc talking machine which was superior in several respects to the $25.00 "Improved" Gramophone, yet could be sold for only $18.00. By the fall of 1899, Seaman had submitted the Valiquet machine for Berliner's approval, and had been given the now-routine refusal. Seaman, having attempted for two years to obtain Berliner's consent to a better Gramophone at a lower cost, finally threw in the towel.

Seaman began marketing the Valiquet machine using his earlier trade name, the "Zonophone." In the spring of 1900, Seaman's Zonophone was licensed under American Graphophone patents, which gave the Columbia forces a "foot in the door" of the disc talking machine market. Additionally, Seaman accepted a consent decree in court, whereby he "admitted" as an agent of the Berliner company that American Graphophone patents were infringed by the Gramophone. American Graphophone quickly obtained an injunction against Berliner, effectively shutting down Gramophone operations. Meanwhile, Columbia dealers began selling Frank Seaman's Zonophone.

Eldridge Johnson, the formerly sheltered sole supplier of Gramophones to the Berliner company, found himself with a new factory full of unsold machines. He was forced to enter the retailing business, with which he had no experience. Leon Douglass, a high-strung talking machine promoter from Chicago, was brought in to help. To the financially-strapped Johnson's dismay, Douglass plowed every cent possible into advertising. Fortunately, Douglass knew his business, and Johnson's sales began to climb. Frank Seaman brought suit, alleging that Johnson's new business in Camden, New Jersey, was merely a front for the Berliner interest. Johnson, however, prevailed in court, and named his new corporation the Victor Talking Machine Company. Emile Berliner resuscitated his Gramophone operations in Canada, destined to watch while Victor, his erstwhile supplier's company, became a global giant.

Frank Seaman, the man who did more than anyone else to promote the nascent disc talking machine, did not reap the rewards of his vision. For several years, Columbia had been waging a campaign to break into the disc talking machine business. The Zonophone was merely a stepping-stone toward that end, and was allowed to wither while a Columbia Disc Graphophone was developed. Seaman's National Gramophone Corporation went into receivership in September 1901. The following month, the first Columbia Disc Graphophones were introduced. Deprived of Columbia's support, Seaman reorganized his Zonophone business, introducing attractive new models throughout 1902 and 1903. However, under virtually constant legal barrage from Victor as well as Columbia, Seaman sold the Universal Talking Machine Manufacturing Company to Victor in September 1903. This transaction was directly linked to the 1903 acquisition of the International Zonophone Company by Gramophone and Typewriter, Limited. Ever the advertising man, Seaman made a pitch to his old adversary, Eldridge Johnson, for the Victor advertising contract. It would have made a fitting and romantic chapter in talking machine history had these two men joined forces, but Johnson was no romantic. He apparently showed Seaman the door.

By the end of 1903, the talking machine business in America had taken on the form it would maintain for the next decade. The "Big Three" dominated the trade, with Edison manufacturing cylinder talking machine products, Victor focused entirely on disc merchandise, and Columbia supplying both types. Pre-recorded cylin-

der records were now being molded in great numbers, rather than laboriously copied or pantographed as before. The new molding process made the large, expensive 5" cylinders obsolete. Disc records had also grown in diameter, from 7" to 10" to 14", finally settling back to 12" as a reasonable compromise of playing time, durability, and ease of handling. The 10" and 12" size 78s would prevail for over half a century.

Sales continued to climb for the "Big Three" companies as the fledgling talking machine business matured into an industry. Its products became more sophisticated, and numerous ancillary businesses developed to supply aftermarket items such as horns and record storage cabinets. The early stages of the talking machine business were evolving into a curious period of refined ebullience. Small workshops were replaced by factories, grimy inventors became ensconced in paneled offices, and the once-crude talking machine was poised to assume the ostentatious garb in which it would achieve enduring acceptance in the homes of millions.

2-24. When the Type "C" Graphophone was introduced in September 1897 to replace the earlier Bell-Tainter derived commercial models, it was necessary to offer a new 6" (long) cylinder for use with it. A correspondingly-named Type "C" cylinder was introduced which conformed to the exterior dimensions of the earlier Type "E" cylinder, and the interior diameter of the Edison-style records. The 6" long format, a vestige of the earliest Graphophones, would persist for decades in office dictation machines. *Courtesy of Jean-Paul Agnard.*

2-25. This one-of-a-kind Type "A" Graphophone from the former Columbia archives suggests the kind of embellishment which could have been offered to improve the appearance of a rather plain-looking machine. The company never adopted such fancy relief motor plates, and future stylistic efforts were concentrated on cabinetry. *Courtesy of the Charles Hummel collections.* (Value code: VR)

2-26. The Empire State Phonograph Company of 76 University Place, New York City, offered this coin-operated Edison Class "M" electric Phonograph in mid-1897. The machine employed a Tewksbury or "Kansas" repeating mechanism, which was manufactured by the United States Phonograph Company of Newark, New Jersey. The cabinet of this coin-op portended the styling which, by 1905, would predominate the public entertainment phonograph field. The bulky, floor-hugging cabinetry of the 1890s would give way to more graceful designs, elevated on slender legs. *Courtesy of Sam Sheena.* (Value code: A)

2-27. When organized in February of 1898, the Universal Talking Machine Company was not a producer of disc talking machines. It would later become well known for the "Zonophone," but in 1898, through an association with Emile Berliner's distributor, the firm functioned in a peripheral capacity, adapting conventional Berliner Gramophones to coin-slot operation. (Value code: VR)

2-28. Opening the all-enclosing cabinet of the coin-op reveals a conventional Berliner "Improved" Gramophone. One is tempted to surmise that Universal was trying to disguise the true nature of the mechanism, but contemporary documents prove that the Berliner Company was cognizant of and consented to the coin-slot conversion. In the words of Berliner Gramophone Company president, Thomas Parvin: "I can find no objection to the manufacture and sale of this type of machine."

2-29. This plate listed operating instructions for the coin-op Gramophone. Note the two vague patent statements at the bottom. The U.S. patent "allowed" was probably No.631,911, filed October 28, 1898 and granted August 29, 1899 to Louis P. Valiquet. Orville LaDow, president of Universal, later testified: "It proved however, impossible to market such machines and only a very few have been sold." (Not surprisingly, since this machine sold for $50.00.)

2-30. The introduction of the "Standard" Phonograph in early 1898 turned the tide in the fortunes of the Edison company. This $20.00 machine would become the most popular model in the Edison line. Soon, the "Standard" in its light oak cabinet could be found in music shops everywhere. At least one merchant attempted to infuse the rather plain cabinetry with some style. In addition to the maroon finish and gilt decorations, the suitcase-style clips were replaced by less conspicuous hooked catches. Two pins in the lid's lower edge aligned with corresponding holes in the cabinet, keeping the latched lid secure. A similar lid arrangement may be seen in the Excelsior talking machine in illustration 2-49. The retailer responsible for these "improvements" remains unknown. (Value code: H)

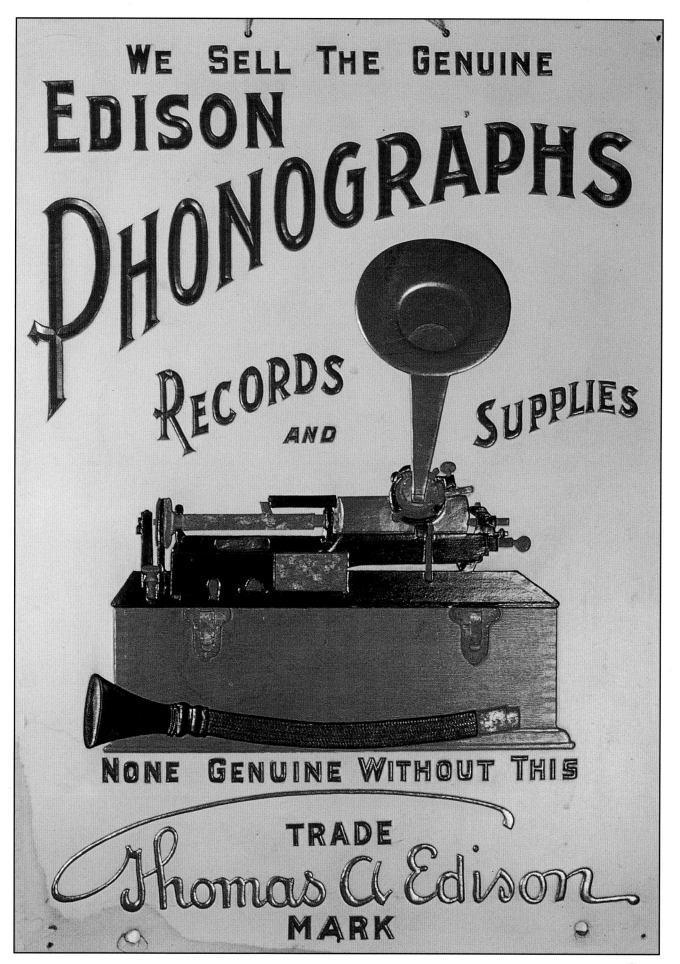

2-32. When Albert T. Armstrong, C.G. Conn, Emory Foster and Joseph W. Jones conspired to manufacture the "Wonder" disc talking machine in November 1897, one of the thorniest problems they encountered was how to provide records for it. The only discs in existence at that time were being made by the Berliner company. Jones, in particular, dreamed of creating a "Wonder" record. He had acquired some technical knowledge of record making while working for Emile Berliner, but discovered that the manufacturing of discs was a prospect fraught with numerous difficulties. Throughout the spring of 1898, the partners struggled to bring the "Wonder" to fruition, and to achieve their goal of producing records. In June, their plans fell apart under the pressure of a suit brought by the Berliner Gramophone Company. Only a small number of "Wonder" machines were produced, and for many years record historians believed that no "Wonder" records had been manufactured. This rare example is evidence to the contrary.

2-33. The desire to develop a practical spring motor for the Edison Phonograph in the early to mid-1890s was embodied in unusual machines from the period, such as this one. This appears at first glance to be an Edison Class "M" electric Phonograph in a custom cabinet of bird's-eye maple, but a winding crank escutcheon on the right hints at the true nature of this machine. It is, in fact, a clever, spring-driven imitation. The designer of the instrument relied heavily upon the cabinet dimensions and basic arrangement of the Class "M" Phonograph. The upper casting was a copy of an Edison, and bore unique decoration. The machine is unmarked, and its origin is unknown. The nickeled mandrel suggests that it was built no earlier than 1898. (Value code: VR)

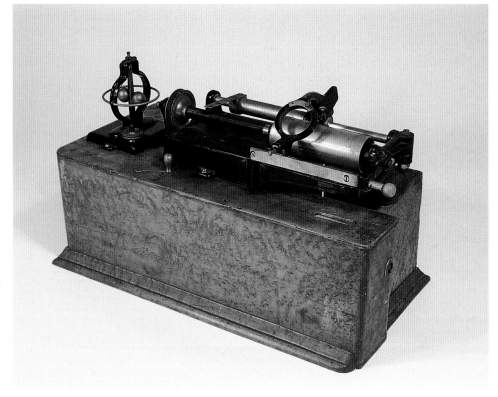

2-34. The motor of the previous example bears a loose resemblance to Edison "Homes" of the period, yet the style of the governor remained similar to that of the Class "M." Aside from the transverse pulley driving the governor, the designer seemed to have contributed few original ideas. The turned legs indicate that the motor was meant to be displayed on occasion. The purpose of the threaded arm on the winding shaft is unknown, but may have been part of a spring tension indicator.

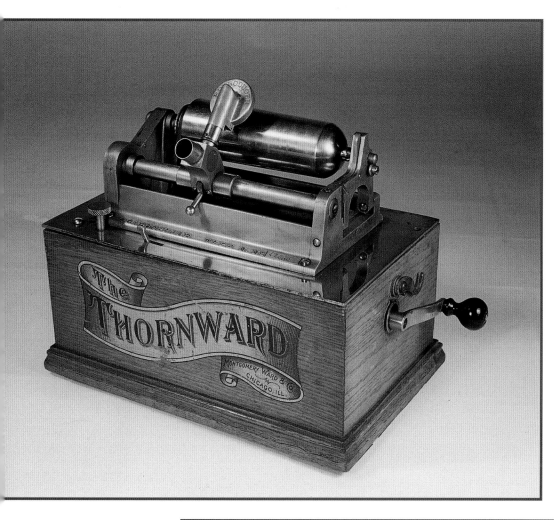

2-35. During the late 1890s, Montgomery Ward, a Chicago mail-order firm, sold a hybrid cylinder talking machine known as the "Thornward." Manufactured by the American Graphophone Company, the machine combined the upper works of a Type "N" Graphophone with the motor of a Type "A." Montgomery Ward described the "Thornward" as "A complete $35.50 outfit for only $22.50 [not including freight]. No such value was ever quoted at this price." Included in the "outfit" was the machine, a dozen cylinder records, a recorder, record carrying case, oilcan and chip brush. The customer could even save 45 cents by sending cash with his order. During its lifetime, which may have stretched as late as 1901, the "Thornward" appeared with two different decals on the front of the cabinet. The banner, shown, is the rarer of the two, the other being a simple rectangle. *Courtesy of Bob and Karyn Sitter.* (Value code: G)

2-36. The "Toy" Graphophone of 1899 was the direct product of the American Graphophone Company's frustrated efforts to enter the disc talking machine market. Unwilling, or unable, to risk introducing a Gramophone look-alike under its own name, American Graphophone and Columbia chose to play their strong suit: a disc novelty based on Columbia's much-vaunted wax recording and reproducing technology. In the foreground are the 3 3/8" diameter brown wax, vertically recorded discs which play from the center outward. (Value code: VR)

2-38. A Graphophone store placard from 1899, measuring 9 3/4" x 12 1/2". *Courtesy of the Sanfilippo collection.*

2-37. The "Toy" was introduced for the 1899 Christmas trade, at $3.00, including one set of five discs. Despite a rapid price reduction to $1.50, the "Toy" faded from the market, and the Graphophone Company commenced a two-year struggle to bring a viable "Disc Graphophone" before the public.

2-39. In 1899, distributor Frank Seaman made repeated attempts to interest the Berliner Gramophone Company in backing alternative designs based upon the company's patents. One model suggested to Berliner by Seaman was this one, built by R.L. Gibson of Philadelphia. By the standards of 1899, this machine was rather advanced: a threaded horizontal crank (note the similarity to future Victor cranks), a turntable capable of playing discs larger than 7" in diameter (although none yet existed), and a brake which acted on the governor (as would be employed later by the Disc Graphophone). (Value code: VR)

2-40. Although it used a standard Berliner turntable, soundbox and traveling arm, much of the attendant hardware of the Gibson was unique. Without the aid of tools, it could be quickly disassembled for storage or travel.

2-41. The motor of the Gibson displayed a simple, yet forward-thinking design. Frank Seaman saw in this machine, as he had in an alternative model designed by Levi Montross in 1897, the opportunity to sell a spring motor Gramophone for less than $25.00. Section Eight of Seaman's contract with the Berliner company allowed him to submit samples of machines of comparable quality if they could be manufactured for at least 5% less than current products. Officers of the Berliner Gramophone Company, however, had covert financial ties with the current supplier of Gramophones, Eldridge R. Johnson. Such entanglements eventually destroyed Berliner's company in the United States, and doomed the barely-realized Gibson to obscurity.

2-42. The Berliner "Improved" Gramophone was modified for coin-operation by the Gramophone Company, Limited in England (and other European Gramophone branches) and introduced at the turn of the century for £8 8 0. Perhaps one of the most elementary coin-slot talking machine designs to enter production, the "Automatic Gramophone" merely prevented winding until a coin was dropped in the slot. The customer selected a disc from the area beneath the horn, and followed the directions: "CHANGE THE NEEDLE: PLACE THE 1p IN SLOT – WIND FULLY, BRING NEEDLE POINT TO OUTER EDGE OF DISC. PRESS KNOB WELL HOME. NEW NEEDLE IS ESSENTIAL FOR EVERY TUNE." Fresh needles were kept in the cup fastened at the upper left.

2-43. Henry Lioret was a French clockmaker who, during the late 1890s, offered cylinder talking machines which were both sophisticated and original. Drawing upon a method of motive power already adapted to the "Tinfoil Phonograph," his "Lioretgraph à poid" ("No. 3") used suspended weights to drive the mechanism. The machine played celluloid cylinder records through an aluminum amplifying horn. Lioret offered a number of far more portable models as well, including a popular talking doll. *Courtesy of Jean-Paul Agnard.* (Value code: VR)

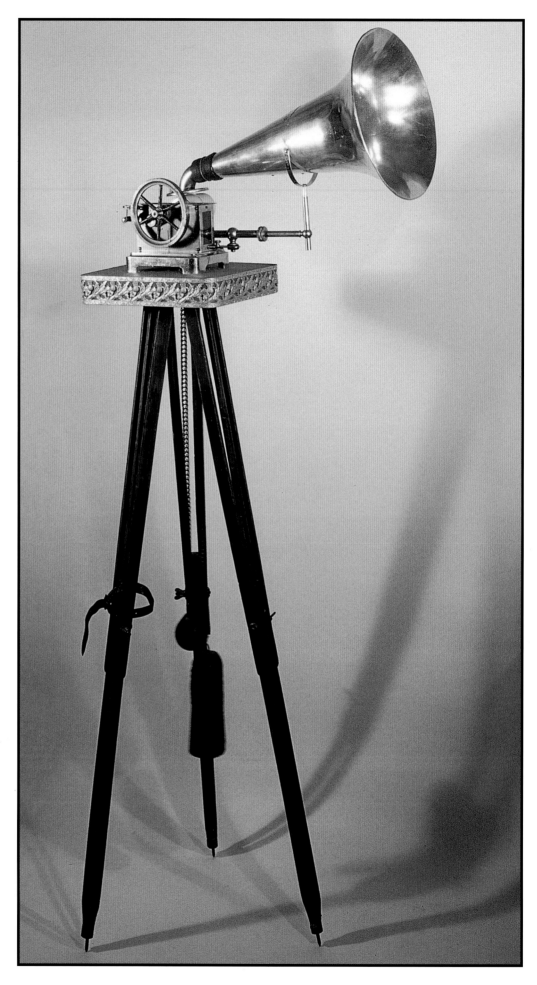

2-44. In 1899, the Girard mail-order firm of France initiated what was to be a lengthy relationship with the talking machine by offering this interesting "Omega" cylinder model. The "Omega" foreshadowed the many, many close imitations of the Columbia Model "B" Graphophone ("Eagle") in "reversible" cabinets that would be manufactured and sold in Europe over the next ten years. Yet, the "Omega" was elevated above the ordinary "knock-off" by the vertical position of its governor. The governor also incorporated a worm drive, something Columbia would not adopt in its motors for a number of years. Girard went on to promote Pathé-manufactured cylinder and disc machines by mail, offering installment terms such as those employed by the Babsons of Chicago to sell Edisons. *Courtesy of Jean-Paul Agnard.* (Value code: G)

2-45. The useful life of the Berliner "Improved" Gramophone, introduced in 1897, was much longer in Europe and Great Britain than it was in the United States. Frank Seaman, Berliner's erstwhile sales agent, had been interested in superceding the "Improved" Gramophone almost from its inception. Although the machine managed to survive reasonably unchanged into the "Eldridge Johnson" period that followed, it was supplanted after 1901 by more mature models such as the "Royal." In Europe, however, the "Improved" Gramophone represented an instrument of proven success. In various incarnations, it remained in the catalogues of Victor's Gramophone cousins into 1904. This distinguished-looking French version in mahogany, circa 1901, was first known as the "No.5" (125 FF), and later as the "Mozart." It sported the nickel-plated brass horn characteristic of European Gramophones. (Value code: D)

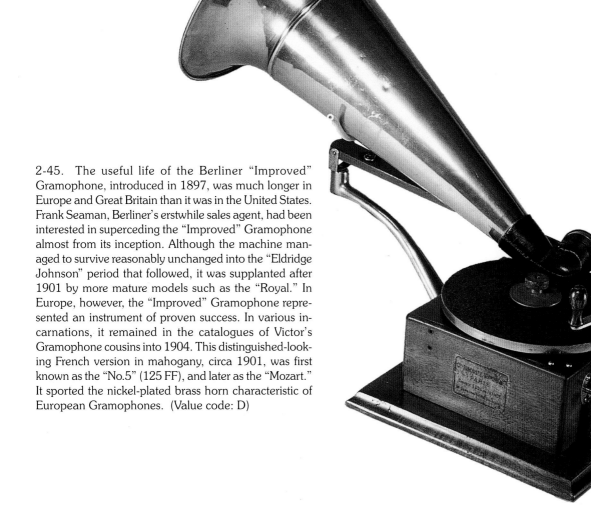

2-46. A highly unusual Berliner "Improved" Gramophone (later style, 1900) with gold-plated hardware. Although gold-plating became increasingly common on higher-priced disc talking machines during the first two decades of the twentieth century, in the 1890s it was all but unheard of. To find a utilitarian machine such as this Gramophone with splendid embellishment certainly indicates a special order. *Courtesy of Gary W. Blizzard.* (Value code: VR)

2-47. The celluloid tag with its attribution to the "Berliner Gramophone Company, Philadelphia, Pa." dates this machine squarely in the few months prior to June 25, 1900. Emile Berliner, deprived of his sales agent Frank Seaman in April of that year, sold Gramophones thus marked in the Philadelphia area until permanently enjoined in late June, when he lost a court case. *Courtesy of Gary W. Blizzard.*

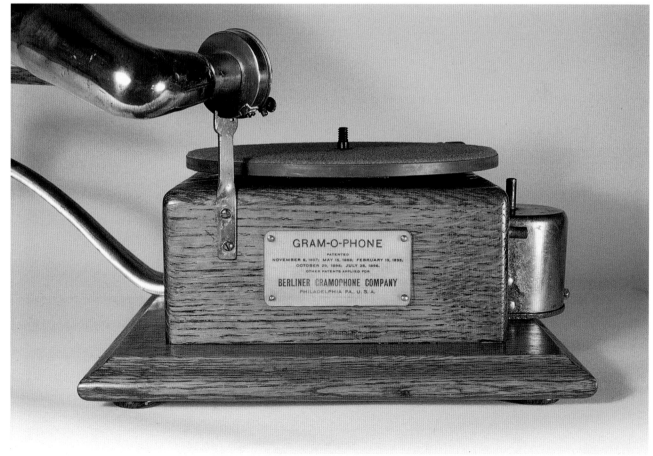

2-48. The only Pathé cylinder machine to borrow from Edison styling was "Le Gaulois" (the Frenchman). Loosely based on the first version of the Edison "Gem," "Le Gaulois" sported a cast iron body for most of its life. However, when it was introduced in 1900, the machine had sides constructed of wood heavily enameled to resemble metal. Here, the first model "Le Gaulois" is fitted with a Pathé glass horn, described by the company as "metallized crystal." *Courtesy of the Domenic DiBernardo collection.* (Value code: E with glass horn)

2-49. In the late 1890s, Germany's Excelsior company got an early start in the cylinder talking machine business by producing a Type "A" Graphophone look-alike. It soon began manufacturing cylinder talking machines of distinctive design, which showed none of the mass-produced conformity that would characterize the machines it sold under a plethora of appellations after 1905. (Value code: F)

2-50. This Excelsior "Modell 1900," designating the year, had charmingly peculiar styling from the delicate connection between reproducer, horn and carriage to the "egg-beater" mandrel. This last feature was most likely present to forestall tapered mandrel patent infringement problems, as it was in other machines such as "Rectorphone" of 1906.

2-51. The complicated *art nouveau* decals which graced the Excelsior "1900" cabinet bear special noting here.

2-52. Augustus Stroh was the builder of the first talking machine in England, and continued his phonographic experiments for much of his life. Although believed to be the designer of the "Tinfoil Phonographs" marketed by the London Stereoscopic Company through the mid-1880s, Stroh's subsequent machines were probably custom-made by him. In any event, Stroh phonographs are infrequently encountered, and each is unique. The example shown employed a playing mechanism consisting of a modified Type "C" Graphophone upper works. Characteristic Stroh features include a large spoked drive pulley, machined brass parts, an electric motor with rheostat, a shifting lever to change threads-per-inch drive, and access doors at the sides of the cabinet. The reproducer carriage has been removed in this view. *Courtesy of Sam Sheena.* (Value code: VR)

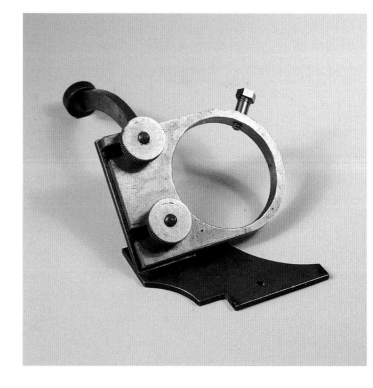

2-53. A close-up of the brass reproducer carriage mounted on its arm. By loosening the two knurled nuts, the carriage may be slid from the arm, and another identical carriage put into place. Stroh phonographs were typically capable of recording and playing at various numbers of threads (grooves) to the inch. Like the Edison Phonographs of 1908-1912, which used different reproducers for two or four-minute records, the Stroh machines needed rapidly interchangeable parts to accommodate various standard and nonstandard cylinders. *Courtesy of Sam Sheena.*

2-54. A view of this Stroh phonograph removed from its cabinet reveals the stock electric motor as used on Graphophones since 1893. Stroh added a rheostat for speed control. Additionally, he incorporated a gear shift mechanism to change the threads-per-inch advance, as evidenced by the brass gears behind the drive belt. Like all of Stroh's work, this machine was a high-quality, precision instrument. *Courtesy of Sam Sheena.*

2-55. Although many, many European firms turned out virtually exact copies of the reliable Columbia Graphophone Type "B" ("Eagle") during the first decade of the twentieth century, at least one French firm imported bona fide "Eagle" mechanisms from the United States and mounted them in domestically-built reversible mahogany cabinets, circa 1900. Shown is a close-up of one such machine. The most interesting thing about it, however, is the locally-made decal. The female figure was an interesting mix: both suggestive of the New World's "Miss Columbia" who adorned Graphophone advertising, and reminiscent of an Old World goddess of plenty. The nomenclature has been cribbed from a genuine Columbia Graphophone of the 1894 period, but misspelled as only a Frenchman with a shaky grasp of English would do: *"perfectionned."* The designation "Type K" must have carried some particular appeal. It appeared again in 1909 on an American disc machine garishly marked "Type K Graphophone." *Courtesy of David Werchen.* (Value code: G)

2-56. Another version of the "Perfectionned Type K." Despite the phony English, this one left no doubt of its European provenance. Beautiful inlay such as this was not something offered to American talking machine buyers, and it's too bad. Collectors today would be happy if cabinets of this quality and charm were more commonly available. (Value code: F)

Opposite page: 2-58. A 10 5/8" x 14" handbill advertising a French talking machine dealer, circa 1900. The Columbia Graphophone Type "B" ("Eagle") had tremendous impact on Europe. Apart from strong sales of American-made "Eagles," a multitude of local manufacturers closely copied the simple mechanism.

2-57. A close-up of the inlaid cabinet of the "Type K."

PHONOGRAPHES & GRAPHOPHONES
BOURRAUX FRÈRES

32-34
Rue Michel-le-Comte

PARIS

MANUFACTURE DE CYLINDRES VIERGES & ENREGISTRÉS

ARTICLES SPÉCIAUX
pour
RÉPARATIONS

USINE
(Seine & Oise)
À VILLIERS SUR MARNE

IMP GREVEL FRÈRES, PARIS

69

2-59. This peculiar "craftsman-made" European cylinder talking machine combined elements of both Phonograph *and* Graphophone design. The motor was small, but the cabinet wide to accommodate the Edison-style tracking mechanism. The reproducer carriage, though reminiscent of Edison, incorporated a Graphophone-type head. The charming floral decoration and red flower horn made this machine a curious object of beauty. *Courtesy of Jean-Paul Agnard.* (Value code: H)

2-60. A stock certificate from the Micro-Phonograph company that Gianni Bettini formed after moving to France following the turn-of-the-century. *Courtesy of the collection of Howard Hazelcorn.*

2-61. In the mid-1890s, Columbia established an early and strong presence in France. By the end of that decade, the announcement "…for the Columbia Phonograph Company of New York and Paris" had been heard on thousands of cylinder records in the United States. Most of the independent cylinder talking machine manufacturing in France would mimic Graphophone design. Of course, Graphophones were easier to copy than more complicated Phonographs, but well-established public recognition of the Graphophone product created a thriving market. Here we see an American-made Columbia "Eagle" in a reversible cabinet, as sold by the firm's Paris branch. The banner decal incorporated the figure of "Miss Columbia," a decal not used in the United States. The *cor-de-chasse* (hunting horn) lends grace to the rather ordinary mechanism. *Courtesy of the Charles Hummel collections.* (Value code: G)

2-62. Gianni Bettini had a genius for invention and a flair for self-promotion. In the mid-to-late 1890s, he positioned himself as the foremost purveyor in the United States of sophisticated cylinder talking machine attachments and records of serious musical content. Based in New York City, Bettini strove to compete with the major forces in the talking machine industry primarily on the strength of his technically advanced "micro-reproducer," which could be adapted either to Phonographs or Graphophones. Just after the turn of the century, Bettini decided to move his operation to France, where a licensing arrangement with Pathé would prove far more successful than his independent efforts. In France, Bettini for the first time began to produce complete cylinder talking machines, such as the "No. 4" shown here. The patented Bettini reproducer, with its protruding counter-weight, can be seen atop a mechanism designed to play either 5" diameter or standard-sized records. *Courtesy of Jean-Paul Agnard.* (Value code: VR)

2-63. Ensconced in Europe, after the turn of the twentieth century, Gianni Bettini enjoyed far more wide-ranging success than he had during the entire decade of the 1890s that he spent in the United States. If imitation is flattery, then Bettini was greatly admired. Pirated versions, as well as licensed copies of his famous "Micro-Attachments" were available in Europe (especially France) and Great Britain. The Bettini knock-off most commonly seen on Phonographs in Britain is the one shown on this Edison "Standard" Model "A," licensed by Edison Bell. Whereas the attachment retained the recognizable elements of a Bettini, its angular shape was unmistakable. *Courtesy of Sam Sheena.* (Value code: D)

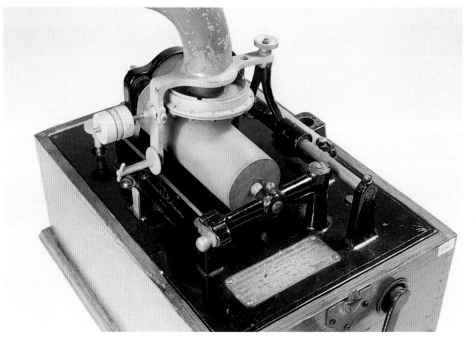

2-64. The Graphophone Type "AF" was one of three machines introduced by the American Graphophone Company in 1901 which were designed to play both standard and 5" diameter cylinders. Most expensive was the "AD," equipped with a six-spring motor and elaborate cabinet for $75.00. Least in cost was the "AB" at $25.00, a showier variation of the "Eagle" Graphophone. The "AF" (shown) was positioned between the others at $50.00, employing the motor and cabinet of the "AG" (an earlier model). The "AF" and the "AD" were the only Graphophones which incorporated two separate drive belts in their mechanisms. (Value code: E)

2-65. The front page of the *St. Louis Grocer and General Merchant*, March 12, 1902, advertising the Columbia "AB." *Courtesy of Mr. R. Chase.*

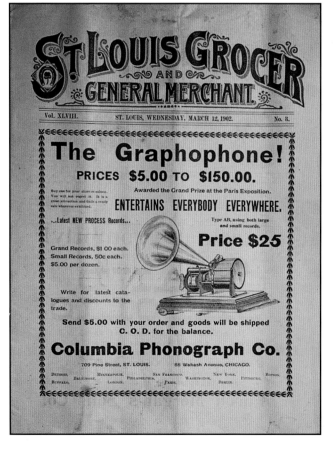

2-66. F.M. Prescott's International Zonophone Company (1900-1903) sold a line of Zonophone disc talking machines in Europe which frequently resembled but were not exact copies of those in the American catalogue. A coin-operated model such as this was never offered in the United States. The cabinet consisted of a Type "C" Zonophone sitting atop a scalloped wooden base which hid the coin drawer. It is interesting to note that this particular Type "C" cabinet (without coin attachment) was available in the United States for a very short period following the bankruptcy of the National Gramophone Corporation in September 1901. *Courtesy of the Domenic DiBernardo collection.* (Value code: VR)

2-67. When the Columbia Phonograph Company began marketing full-size Disc Graphophones in late 1901, two models were offered. The Type "AH" was the larger, while the Type "AJ," (shown) with its 7" turntable, single spring, and 16" horn, was the smaller. The vertical crank was a throwback to the Berliner "Improved" Gramophone, and the "AJ" was the only spring-driven Disc Graphophone to use that archaic design. This very early example featured an unusual speed control and on/off lever. Several varieties exist of the early "AJ," suggesting ongoing experimentation and production modification. *Courtesy of Donald Walls.* (Value code: G)

2-68. Columbia was accustomed to taking the lead when it came to talking machine trends. Yet, it tried for years to break into the disc talking machine business, to little avail. Beginning in 1898 and culminating with the introduction of two Disc Graphophones in October 1901 ("AJ" and "AH"), Columbia tried by every means possible to get a slice of the disc pie. When the Disc Graphophones finally arrived, they were substantially derivative of the Victors and Zonophones which had preceded them by more than a year. One Columbia feature that was distinctive was the use of decoratively cast aluminum arms in the horn assembly. In a single instance, however, Columbia (briefly) imitated the type of pivoted wood arm commonly associated with Victor. This "AK" Disc Graphophone from 1902 sports this rare arm, emblazoned with the Columbia name. (Value code: G)

2-69. The Manhattan Phonograph Company of New York City produced this coin-operated machine in 1901-1902. The instrument was based exclusively on the earliest Edison "Standard" mechanism, which was discontinued in early 1901. The timing suggests that the Manhattan Phonograph Company bought the inventory of obsolescent mechanisms from Edison, or that the company was actually created to dispose of these outdated machines. In any event, the enterprise was short-lived, and no Manhattan coin-ops are known with works other than those of the earliest "Standard." The Manhattan had the distinction of being the only coin-operated talking machine whose chute (upper right) sent the coin through the air (above the cylinder) to be caught by a metal cup (behind reproducer) and deposited below. Such penny showmanship offered a visual attraction not found in other coin-operated talking machines of the period. (Value code: VR)

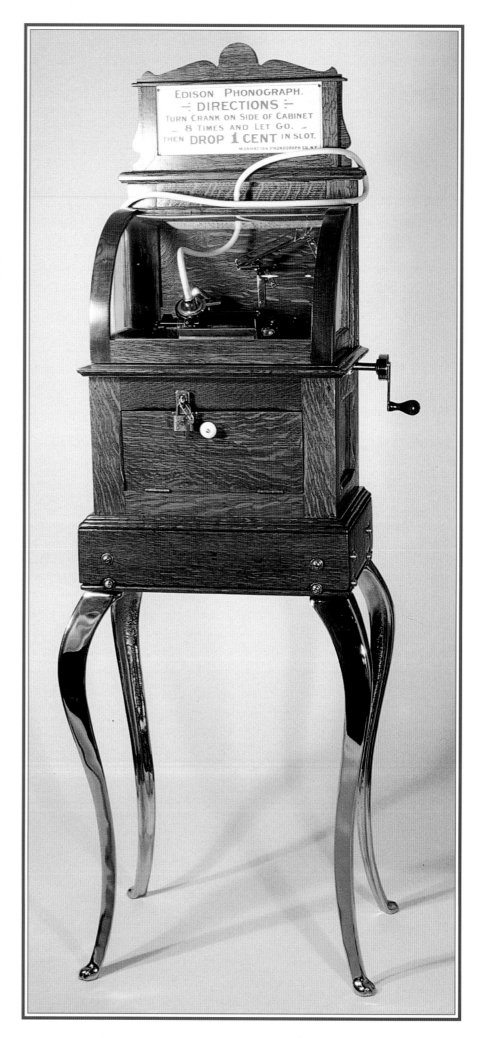

2-70. Two "Manhattan" coin-operated Phonographs in a single oak cabinet. This arrangement simplified the retrieval of coins or tokens, and gave patrons a choice of two records. Note the graceful Queen Anne legs are not metal but wood, unusual for "Manhattan" cabinets. *Courtesy of the Charles Hummel collections.* (Value code: VR)

2-71. The British Edison Bell company (for most of its life Edison Bell Consolidated Phonograph Company, Limited) had a long and eventful existence, mostly under the stewardship of James Hough. The firm frequently struck a chauvinistic note in its advertising, touting "British made… British artists…" and other such appeals to national pride. It is interesting to note, therefore, that a briefly-offered line of Edison Bell cylinder talking machines which included this "Empire" of 1902 was clearly contracted with a German manufacturer. The fact is, Edison Bell could do pretty much as it wished until the middle of 1903, when the Edison and Bell-Tainter patents which it had controlled for ten years finally expired. The company even imported a Pathé "Coq" cylinder machine which was offered as an Edison Bell. *Courtesy of Jean-Paul Agnard.* (Value code: H)

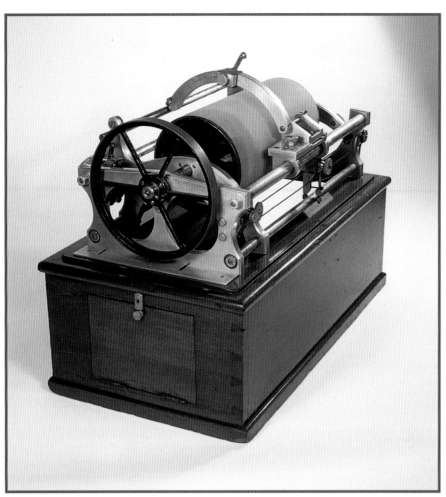

2-72. Another Stroh phonograph, this one designed from the ground up by Augustus Stroh. The dovetailed cabinet carried his signature access doors, and the upper works were constructed from aluminum. The large spoked drive pulley was present, in addition to a gigantic 10" long mandrel of 5" diameter (of the Pathé "Le Céleste" type). The March 4, 1903 issue of *Cassell's Popular Science* featured an article titled "The Phonograph," which discussed "...Mr. Stroh, who has taken so much active interest in the phonograph since its first inception by Mr. Edison." The article illustrated several of Stroh's phonographs. "He has made many of these machines, each one becoming obsolete almost as soon as it was completed." A machine was shown similar to that pictured here, with the explanation "...five-inch records, with of course a correspondingly extended surface, are coming into vogue. The extra surface is, however, not used to take longer records, but it is invaluable, seeing that it allows greater space for each sound wave, thereby securing greater detail." As we have seen from Stroh's earlier work, this principle was not new to him. The reproducer is missing from this example. *Courtesy of the Domenic DiBernardo collection.* (Value code: VR)

2-73. A close-up of the previous machine, showing the attribution to "A. Stroh, London." The gear change lever, which accommodated 100, 150, and 200 threads-per-inch recordings, can be seen at the upper right. In 1901, the then 73-year-old Augustus Stroh patented a violin specifically designed for the recording industry that became known as the "Stroh Violin." Stroh's son, Charles, assumed control of the violin business, leaving Augustus free to pursue the design and construction of fine talking machines such as this one. *Courtesy of the Domenic DiBernardo collection.*

2-74. Elaborate decoration was more often used on European disc talking machines, especially German, than the prosaically-styled instruments sold in America. However, Victor and Talkophone both flirted with high ornamentation in such little-seen machines as the Victor "Deluxe Monarch" and Talkophone "Sousa." The American Zonophone shown here sports a novel cabinet for which there is no documentation other than a registered design. The Renaissance style, with four intricately detailed claw feet, places this machine high in the empty ranks of richly or imaginatively decorated American external horn disc talking machines. *Courtesy of the Domenic DiBernardo collection.* (Value code: VR)

2-75. The person in the Pathé research and development department who signed off on the company's mechanical fan ("Eventail Automatique," 37.50 FF in 1903) was undoubtedly fired. The product proved a failure, and the unsold fans were refitted as rather bizarre-looking cylinder talking machines. A mandrel was affixed where the fan's central plume had discreetly oscillated. Somehow, the simple aluminum horn seems to clash with the otherwise baroque styling. *Courtesy of Jean-Paul Agnard.* (Value code: VR)

2-76. An extremely interesting member of Bettini's catalogue from the French period was the "No.8." The Bettini company used this designation to identify at least two very different, successively-produced cylinder models. Evidence suggests the "No.8"s were produced under contract by the same firm which sold the popular "Phénix" cylinder machines in France. Even though the example shown is an undisguised copy of an American Graphophone Type "B," the stamped rectangular base-plate, exposed feed-screw and characteristically-shaped winding key betray its kinship with "Phénix." The horn is typical of the kind sold with smaller Bettini machines: the body and segmented elbow are steel, painted Chinese red, and the bell is polished aluminum. *Courtesy of the Domenic DiBernardo collection.* (Value code: D)

2-77. The Victor "Royal" began life in 1901 as the smallest "front-mount" machine in the line at $15.00. With the appearance of the revolutionary Victor "Tubular (or 'Rigid') Arm" in October 1902, the "Royal," like its larger siblings, became available so equipped. Pictured is an example bearing an unusual trademark decal on the front of the cabinet, as well as a dealer's label. *Courtesy of Bob and Karyn Sitter.* (Value code: E)

2-78. The Victor Talking Machine Company made rapid improvements and modifications of its disc models in the 1901-1904 period. This created leftover inventories of obsolescent parts, tools and dies. One solution to the problems created by accelerated development was seen in the series of Victor "Premium" machines. These patchwork "Premiums" used up older parts and kept existing tools and dies working. Another response to the same situation involved Victor's European "cousins": the network of Gramophone companies in nations like France, Great Britain and Germany. In Europe, certain American-made parts, motors and tone arms persisted considerably after their discontinuation in the United States. The European Gramophone branches imported Victor parts and used them to create machines which resembled, but were fascinatingly different from, instruments in the ordinary American catalogue. One such model is pictured here. Ostensibly it is a Victor "Monarch Special," circa 1902, with triple-spring motor, "rigid arm" attachment and Concert soundbox. In fact, the motor is identical to the single-spring model used in the so-called "Johnson Monarch" (discontinued after 1901). Like the "Johnson Monarch," the Compagnie Francaise "Monarch" shown here does not employ a hinged cabinet lid. Whereas the "rigid arm" attachment was identical to the ones used in the United States, the horn was of domestic origin, nickeled in the European fashion. Furthermore, machines equipped with the "rigid arm," offered only briefly in America, were sold in Europe for about two years. *Courtesy of David Werchen.* (Value code: E)

2-79. These two machines were exhibited by the American Graphophone Company as a system to record, reproduce and shave "mail-able" cylinders. The shaver (left) and the Graphophone (right) both featured small-diameter mandrels for specially-sized cylinders. Note the 10" nickeled horn for size comparison. The shaver is marked "Model 1005." The Graphophone carried a plate that reads in part "Exhibit No.352." It is believed that this system of mailing cylinder correspondence was never put into commercial use. *Courtesy of the Charles Hummel collections.* (Value code: VR)

2-80. In the realm of miniature "mail-able" cylinders for correspondence, one earlier effort bears mentioning. Illustrated is the British "postal cylinder" of 1891, a feature of the Edison Bell "Commercial" electric Phonograph. This narrow diameter, 4" long wax-coated cylinder fit over the main arbor of the machine after the mandrel had been removed. Edison Bell imagined a time when these little cylinders would replace brief written correspondence, but the idea did not catch on. Shown is the "postal cylinder" and its pasteboard mailing tube, with wooden peg insert. *Courtesy of Sam Sheena.*

2-81. Pathé's "Le Gaulois" was the sincerest form of flattery. It copied the first model of the Edison "Gem" Phonograph, except for the inclusion of a Graphophone-style carriage and reproducer. "Le Gaulois" came in a variety of colors including black, dark green, red, gray, and shades of blue. Here we see the machine sporting an unusual aluminum silver finish, with red details. "Le Gaulois" was sold by mail order under the name "Le Ménestrel" (in blue). The silver-finished version was marked "Le Haut Parleur" (the loud speaker). (Value code: F)

During the period 1904-1911, a prodigious amount of activity occurred in the European and British phonograph industries. In fact, so many and varied were the models produced, that the plethora of styles and the difficulty of properly documenting them has somewhat discouraged appreciation of the entire category of European instruments.

Today, European talking machines occupy an under-rated area of the antique phonograph hobby. In the United States and Canada, they are not generally recognized or understood, although examples can be found in a great many collections. Even in Europe, where machines of domestic origin are enthusiastically collected, the highest emphasis is frequently placed on certain well-known American instruments.

The trajectories of the European and American talking machine industries during the first decade of the twentieth century were essentially reversed. In the United States, the trend was toward reduction to, and glorification of, a small number of popular models. In Europe, great diversity remained undiminished throughout the period. To better understand the differences between American and European machines, let us examine the development of the respective environments from which they emerged.

SMALL ENTERPRISE BOOMS AND THEN BUSTS IN THE UNITED STATES

In America, around 1904, a great blossoming of activity occurred in the talking machine field. Independent manufacturers and distributors sprang up in a dizzyingly expanding market. Larger players such as Victor, Columbia and Edison had not completely consolidated their power. Concerning cylinder talking machines, Edison and Columbia had cross-licensed each other in 1896. However, in the disc talking machine business Victor and Columbia had only recently cross-licensed each other— in December of 1903. For this reason, these two well-capitalized disc talking machine companies had not yet coordinated their attacks on the competition. The disc

field, then, was uncluttered by the moldering corpses of "independents" which already littered the cylinder arena. Into this fresh disc market little companies optimistically plunged. The fates of the smaller firms were often interrelated, and the success of one could mean good luck for a number of others. A case in point was the O'Neill-James Company of Chicago.

In 1904, soon after it was founded by Arthur J. O'Neill, a former traveling salesman and advertising professional, the O'Neill-James Company began marketing a little cylinder talking machine called the "Busy Bee." The machine was in actuality a Columbia Type "Q" Graphophone, manufactured by the American Graphophone Company of Bridgeport, Connecticut. There was, however, a gimmick behind the "Busy Bee" that distinguished it from the identical machine sold under the Columbia name. The "Busy Bee" was equipped with a mandrel very slightly larger than the one Columbia (or Edison) used. Hence, the customer had to buy special "Busy Bee" cylinder records (also manufactured by American Graphophone) to fit it. A clever scheme, indeed, and it proved so successful that O'Neill-James was encouraged to introduce a disc talking machine. To do this, the little company left the patent protection of Columbia's umbrella.

In 1906, O'Neill-James contracted with the Hawthorne and Sheble Manufacturing Company of Philadelphia to construct disc talking machines under the "Busy Bee" name. Again, O'Neill-James enforced the sale of their own records by affixing a protruding lug to the "Busy Bee" disc machine's turntable, which allowed only specially-prepared "Busy Bee" discs to be played. Hawthorne and Sheble, the contract manufacturer and another striving independent, was then approaching the peak of its success as a producer of talking machines, accessories, horns and record storage cabinets. The time would come, however, when industry giants such as Victor would turn their attention to the retail revenue being siphoned off by H&S and other budding entrepreneurs. This attention, once focused, would wither small companies like violets in the summer sun.

Wholesale Dept. on Disc Talking Machines.

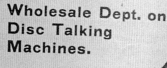

GRAND BUSY-BEE DISC

Never in the history of the manufacture of the Disc Talking Machine has there been a Machine of the high quality, as well as appearance. It is equipped with the Morning Glory Horn beautifully enameled in red, has Moberly's wonderful Sound Box, weathered oak Cabinet the Motor and Gear having all tempered steel bearings. In fact there has never been a Machine offered to the public like it for the money. Price............$18.00

We are Jobbers for Columbia Graphophones and Supplies.

We can furnish you with Columbia Records in Disc— XP Cylinder or BC Twentieth Century (half foot long). The latter contain nearly twice as much of a song or instrumental piece as the ordinary Cylinder Record; and are usable on Columbia machines as follows:

Type BF, BG, BM and BC. (See description of these machines.)

The Following Shows a Few Types of our Popular Disc Graphophones.

ALUMINUM TONE ARM GRAPHOPHONE.
Columbia Majestic, Type BD, $100.00

The very best disc talking machine ever constructed. Aluminum tone arm. Powerful spring motor. Will run ten or more records at one winding. Can be wound while playing. Solid mahogany cabinet, highly polished, piano finish. Twelve inch turn-table encircled with an ornamental aluminum rim. Columbia analyzing reproducer. Automatic needle clamp, doing away with thumbscrew. Nickel silver finish floral horn 22½ inches long, 23½-inch bell. 200 needles. Two-part needle box for new and used needles.

Type BD, Majestic Graphophone. This instrument is the peer of all disc talking machines and represents the highest achievements in the art of Graphophone construction. It is unapproachable at any price.

Above machine equipped with motor to run three records, $75.00.

When wanted, order: Columbia Imperial, Type BJ.

Further on you will see Type BH and should you have any other Type of disc machines on your mind we will be pleased to give you full particulars on any type manufactured.

For descriptions of types not already shown see our Instrument pages.

The Lyric Reproducer.

Through this the human voice is produced (with all the sweets and tone of nature.

Wholesale Dept. on Cylinder Machines.
QUEEN BUSY-BEE GRAPHOPHONE

This machine has all the latest improvements and is equipped with the Lyric reproducer containing a genuine sapphire. Oak cabinet with oak carrying cover. Beautiful Morning Glory Horn making it a wonder in itself.

You can make records at home with this machine which is one of the great amusements of the talking machine.

LYRIC REPRODUCER GRAPHOPHONES

In selecting a talking machine you should consider first of all **the Reproducer.** Upon this all of the pleasure which you anticipate depends. In the improved Lyric Tone Reproducers, with which 1906 model Graphophones are equipped, an entirely new scientific idea in the reproduction of sound is introduced, resulting in a vastly improved reproduction. If you have not heard **"THE NEW LYRIC TONE"** you have no idea how perfect a reproduction you can obtain on the Lyric Reproducer Graphophones. Remember, in the greatest world contests, **all manufacturers being represented,** the Graphophone has always taken first and highest possible awards.

Grand Prix, Paris, 1900; Double Grand Prize, St. Louis, 1904; Highest Award, Portland, 1905.

3-1. A 1906 O'Neill-James catalogue. At the top left is the Hawthorne and Sheble-built Busy Bee "Grand" disc machine. H&S offered a similar instrument under it own Star brand. On the right can be seen the "Queen" Busy Bee, the better of the two Columbia-built cylinder models the company offered. O'Neill-James also sold unmodified Columbia cylinder and disc Graphophones.

The O'Neill-James Company had become linked to the success of Hawthorne and Sheble. Furthermore, "Busy Bee" disc records were being pressed by independent operators such as Leeds and Catlin, and the American Record Company (related to H&S). A network of small enterprise had arisen. Unfortunately, it was based more on zeal than on satisfactory patent protection.

In the disc talking machine industry, Victor and Columbia, through cross-licensing, shared Victor's so-called "Berliner Patent" (No.534,543, February 19, 1895) and Columbia's so-called "Jones Patent" (No.688,739, December 10, 1901). The former related to fundamental disc talking machine design, and the latter to the manufacture of lateral cut disc records. Victor and Columbia litigated in the name of their respective patents, but were immune (in the areas covered by their agreements) to each other's legal actions. During 1909, Victor was prosecuting Hawthorne and Sheble for patent infringement, and also sued O'Neill-James (and its sister-firm, the Aretino Company). Around the same time, both Columbia and Victor took Leeds and Catlin to court. Other independent disc manufacturers were also sued. The web of mutual welfare that linked the small firms began to collapse.

By 1911, nearly all the independents in the American talking machine and record industry had been crushed. Interestingly, though H&S and Leeds and Catlin succumbed, O'Neill-James survived. The little Chicago company that had begun under Columbia's patent protection returned to the shadow of Columbia's aegis. Arthur J. O'Neill subsequently consolidated O'Neill-James with his other talking machine enterprise, the Aretino Company, which operated through 1913.

This brief history illustrates the redistribution of authority that took place in the United States phonograph industry during the period 1904-1911. A pattern of expansion among the principal forces and diminution of the challengers temporarily choked the variety out of the American market. In Europe, however, similar compression did not occur.

3-2. Hawthorne and Sheble Manufacturing Company of Philadelphia introduced a line of Star external-horn disc talking machines in July 1907. Several interesting features distinguished these instruments. Thomas Kraemer's "Yielding Pressure Feed," a device installed within H&S tone arms, facilitated the passage of the soundbox across the record. The soundbox diaphragm could be replaced without tools. Furthermore, the larger Star models were equipped with receptacles for the disposal of used needles. In Chicago, Arthur J. O'Neill sold H&S disc machines under the "Busy Bee"/"Yankee Prince" and "Aretino" brands. O'Neill's machines were modified to accept specially-prepared records. Only Aretino discs, with their 3" diameter center hole, would fit the unusual turntable of the 1908 Aretino shown here. (Value code: G)

3-4. In the archeology of the talking machine, as in the more well-known bones and mummy variety, intriguing clues help us gain a real sense of historic events. Arthur J. O'Neill was enjoined from selling the infringing *mechanics* of Hawthorne and Sheble machines. However, there was no law against recycling the H&S cabinets for use with Columbia mechanisms (Columbia and Victor were cross-licensed, and thereby largely immune to infringement). To do this required a little work, since banner decals already had been applied to the cabinets, and holes for the H&S motor controls already had been drilled in the wood. Furthermore, the modifications needed to be done hurriedly: O'Neill's business was at a standstill. An empty H&S cabinet was reoriented by 90 degrees, with the former front panel becoming the right side panel. A thin oak "patch" was placed over the existing H&S motor control holes so that a Columbia back-bracket, tone arm and horn would largely disguise the alteration. The red banner transfer carrying the Aretino name was then re-moved and reapplied to what had become the "front" panel of the cabinet. In this illustration, we see the thin board upon which the Columbia bracket was mounted in order to hide the modification of the cabinet. We may even discern a few shiny fragments left by the hasty removal of the decal from its original location on the left-hand panel. *Courtesy of Michael Sims.*

3-3. 1908 and 1909
were bad years for independents
in the talking machine trade. In the midst
of a business depression, Victor and Columbia were busy slugging away at the little guys, lighting one struggling entrepreneur after another the way to dusty death. Arthur J. O'Neill's Aretino Company of Chicago sold disc talking machines produced by independent manufacturer Hawthorne and Sheble of Philadelphia. Victor sued H&S for making patent-infringing machines, and sued Aretino for distributing them as "Aretinos." By the end of 1909, both of O'Neill's companies, Aretino and O'Neill-James ("Busy Bee" and "Yankee Prince"), were prevented from selling any machines or any records manufactured by other independents. Arthur J., left without a means to continue his livelihood, went to Camden, hoping his "old friend" Belford Royal (one of Eldridge Johnson's long-time associates) could put in a good word for him. Unfortunately, the word from the top was that O'Neill was *persona non grata.*' O'Neill left in defeat and traveled to Bridgeport, Connecticut, to meet with Columbia. The "Aretino" machine shown here was the result of this trip. *Courtesy of Michael Sims.* (Value code: G)

3-5. An interior view of the "transitional" Aretino tells us still more. Columbia motors required more clearance than H&S motors, so a recessed area has been routed out of the base board to prevent the Columbia spring-barrel from bottoming out. To the left is the typical paper maintenance notice that accompanied all Columbia-motored machines. At the rear, the original H&S control holes can be seen. By carefully analyzing this machine, we begin to imagine the hectic work which sparked anew the Aretino Company and gave its brief candle another four years of life. *Courtesy of Michael Sims.*

3-6. In 1905, a company was formed in Lincoln, Nebraska to manufacture double-horned cylinder and disc talking machines under the name "Duplexaphone." The cylinder model was never produced, but the disc machine appeared as the Duplex the following year. Over the next four years, The Duplex Phonograph Company strove to sell its unusual instrument for $29.85 by direct mail, while fending off charges of infringement brought by Victor. Victor finally prevented Duplex from doing business, and by the end of 1910 the affairs of the Nebraska firm had all but terminated. The story, however, continued. Mr. C.Q. DeFrance circulated at letter dated "winter of 1910-11" announcing that he was the new owner of the Duplex company, and was about to release a mechanical-feed model which would not infringe Victor patents. Duplex, in its earlier incarnation, had already unsuccessfully attempted a putative mechanical feed. The letter, couched in understandably partisan language, read, "Friends of the Duplex Phonograph have now the opportunity in some measure to right the wrong inflicted by the talking machine trust, in its nearly successful attempt to crush competition. The old style Duplex… was superior in size, finish and especially tone quality, to the $40, $50, $60, and $75 machines put out by the trust; and being unable to compete as manufacturers, the trust, by skillful manipulation of the patent laws, by the use of unlimited money for law expenses, and by long harrassing [sic] litigation finally did crush the former Duplex Phonograph Company. But the trust did not kill the Duplex Phonograph. It is alive and livelier than ever—a better machine in every way." No evidence has arisen, however, to indicate that the "progressive mechanical feed" Duplex was anything more than an experiment. In 1914, a letter from *Mrs.* C.Q. DeFrance on old Duplex company stationery indicated that she was primarily in the business of providing parts and repairs for existing Duplex machines. She further stated that only 20 "new" Duplex phonographs (presumably the mechanical-feed model) were left in stock, and that no more would be built. "Price $30.00, cash, f.o.b. Kalamazoo, without records." She was selling Columbia disc records, and she offered to accept old Duplexes for exchange against the Columbia Grafonola "Leader" ($75.00 retail) and $55.00 cash (customer to pay the shipping). The purpose of this offer was apparently to obtain a source of parts to keep the repair business active. The last line of the letter, however, suggested the end of a story that began nearly ten years earlier was near, "…this is probably the last year you can secure Duplex repairs…" *Courtesy of the Sanfilippo collection.* (Value code: D)

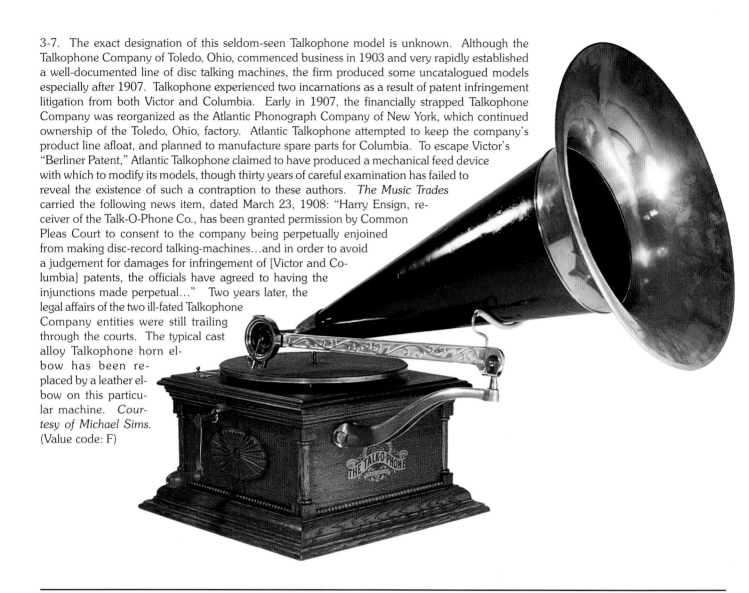

3-7. The exact designation of this seldom-seen Talkophone model is unknown. Although the Talkophone Company of Toledo, Ohio, commenced business in 1903 and very rapidly established a well-documented line of disc talking machines, the firm produced some uncatalogued models especially after 1907. Talkophone experienced two incarnations as a result of patent infringement litigation from both Victor and Columbia. Early in 1907, the financially strapped Talkophone Company was reorganized as the Atlantic Phonograph Company of New York, which continued ownership of the Toledo, Ohio, factory. Atlantic Talkophone attempted to keep the company's product line afloat, and planned to manufacture spare parts for Columbia. To escape Victor's "Berliner Patent," Atlantic Talkophone claimed to have produced a mechanical feed device with which to modify its models, though thirty years of careful examination has failed to reveal the existence of such a contraption to these authors. *The Music Trades* carried the following news item, dated March 23, 1908: "Harry Ensign, receiver of the Talk-O-Phone Co., has been granted permission by Common Pleas Court to consent to the company being perpetually enjoined from making disc-record talking-machines…and in order to avoid a judgement for damages for infringement of [Victor and Columbia] patents, the officials have agreed to having the injunctions made perpetual…" Two years later, the legal affairs of the two ill-fated Talkophone Company entities were still trailing through the courts. The typical cast alloy Talkophone horn elbow has been replaced by a leather elbow on this particular machine. *Courtesy of Michael Sims.* (Value code: F)

ELEMENTS OF THE EUROPEAN MARKET

Although major forces influenced the development of the European phonograph industry during the first decade of the twentieth century, a stultifying patent environment did not exist, as it did in the United States, to prevent the proliferation of minor enterprise. In Great Britain, domestic talking machine production was concentrated in large firms such as the Gramophone Company, Limited and the Edison Bell Consolidated Phonograph Company, Limited. Both of these firms had their start in the 1890s as factors or outlets for American-made talking machines. By 1904, both were well established and had initiated their own products. However, a huge number of European exports flooded English shores, making the market dense with diversity. F.M. Prescott's International Zonophone Company, established in 1900 in Berlin, vied with the Gramophone factions for supremacy in the British and European markets during the

first few years of the century. Then, in June of 1903, the Gramophone Company (at the time called Gramophone and Typewriter, Limited) took over the affairs of International Zonophone. The Zonophone was subsequently re-inaugurated as a budget-priced Gramophone adjunct.

The British Lambert Company sold European-made cylinder talking machines under its mark (as Edison Bell had done, before establishing substantial home production). The wildly industrious German Excelsior company manufactured a host of cylinder models, including "Angelica," "La Favorita," and "New Century," specifically for distribution in the British Isles. Pathé of France opened a London headquarters in 1903 to promote the sale of French-made instruments. In 1903, Edison's National Phonograph Company, having broken with Edison Bell and all previous distribution arrangements, established a British branch. The National Phonograph Company, Limited imported a large number of American Phonographs (primarily the "Gem" and "Standard" models) until it was interrupted by the First World War.

A raft of small English distributors vigorously disseminated various versions of the simple and inexpensive "Puck" or "Lyra" cylinder machine. Georges Carette and Company of Nuremburg, Germany, manufactured the humble "Puck" in over 30 variations concerning horn, shape of the base casting, type of decoration and strength of motor. Carette and the other producers of "Pucks" maintained strong sales links to the British Isles. The popularization in Britain of the Edison four-minute (in playing time) Blue Amberol cylinder record (released in Britain, March 1913) would bring an end to the trade in cheap cylinder talking machines of the obsolete two-minute variety. Any vestiges of "Puck" manufacturing in Germany were erased by the First World War.

In Germany, the manufacture of cylinder and disc talking machines continued robustly through the period 1904-1911. At the turn of the century, Deutsche Grammophon, the Victor Talking Machine Company's German cousin, got an early foothold in the local market as the agent for Berliner's Gramophone. The company would remain a significant force in the German disc industry throughout the acoustic period, manufacturing machines with features distinctive of the Victor/Gramophone family. The International Zonophone Company, even after its take-over by Gramophone and Typewriter, continued to produce a wide range of instruments. The firm's 1905 catalogue listed ten external horn models: eight front-mounted and two back-mounted. Meanwhile, production of "independent" German disc instruments expanded, especially 1907-1909.

A brief review of German disc machine brands traded during the 1904-1911 period results in the following partial listing: Adler, whose 1905-06 catalogue listed ingenious (cylinder and) disc machines, included one which "borrowed" the "Monarch" appellation from the Gramophone; Excelsior, a predictably diverse disc line; Hymnophon(e), external and internal-horn disc machines and coin-ops; Klingsor, a brand with strong British associations, whose gimmick was to place a small harp with tuned strings over the mouth of the internal horn; Neophone, vertical cut machines and discs largely intended for the British market; Odeon, popular and varied disc machines; Orchestrophone, whose 1907 catalogue listed 27 external-horn models and one floor-standing coin-op; Parlophon(e), another brand with ties to Britain; Pionier, whose 1908 catalogue sampling included six external-horn disc machines and five coin-slot models; Polyphon(e), music box maven and talking machine dabbler, which for a time constructed Klingsors; Rena, which catalogued in 1909 nine external-horn models, twelve "Serena" internal-horn models and a few "loud-speaking" machines, intended for the British market. Although France and Switzerland produced talking machines in abundance, of all the European countries Germany proved the most prodigious in churning out countless, often unidentified models in both cylinder and disc

formats. These included phonographs imitating furniture, shaped like Zeppelins, equipped with monstrously large horns, decorated with elves, cherubs, sylphs, heroic friezes, and even disguised as cottages out of the Brothers Grimm.

In December 1906, American expatriate mover-and-shaker Frederick Prescott, who was at this point the head of the International Talking Machine Company of Berlin, discussed the overall European record industry (and its implications in the manufacture of talking machines) with the *Talking Machine World* in New York:

> We are working under no restrictions in Europe: that is to say, there are no fundamental patents, only constructive ones. All you need is a knowledge of the process of manufacturing, with sufficient capital, and then you can go ahead entirely free to produce and dispose of your output....There are about 29 manufacturers of disc records in Germany alone, but the majority are small concerns, and not heavily capitalized....Were it not for the Berliner patent my company would manufacture its goods here [America] and get the same prices we command in any part of the globe, strictly on quality and repertoire....I believe Europe is further advanced in many respects in the talking machine line. This is because there is not so much patent restriction, and the competition is much freer to develop and make improvements than in this country [the United States].

In France, the Columbia Phonograph Company's Paris office was very busy during the late 1890s, importing cylinder Graphophones from America while Edison was still attempting to develop a mature line of Phonographs. The Pathé brothers, Charles and Emile, commenced their talking machine business at this same time. They first distributed, then imitated Columbia Graphophones. Even as Pathé cylinder machines evolved independently, they continued to reflect significant elements of Columbia design —except for "Le Gaulois" which resembled the first version of the Edison "Gem." All other brands of cylinder machine ran a distant second to Pathé in France. Clockmaker Henri Lioret sold his delicate and precise mechanisms, Gianni Bettini sponsored a line of instruments based on his patented reproducer technology, Maison de La Bonne Presse, Dutreih and others contributed a wide variety of cylinder models—but Pathé remained unshakably at the top.

In the French disc machine field, Gramophones from the local affiliate, Compagnie Francaise du Gramophone, and Zonophones from International Zonophone entered the market at the turn of the century. Cylinder talking machines, however, remained dominant until Pathé introduced disc instruments based on the alternative "vertical cut" system in 1906. Pathé discs were played by a sapphire ball stylus rather than a steel needle. Despite heavy competition from the Gramophone, Zonophone and an array of steel-needle disc machines from Germany and Switzerland, Pathé popularized the vertical cut system and encouraged its brief adoption in the United States. One French brand that (fleetingly) challenged Pathé in the disc market was the "Pantophone," sold by A. Morhange, Paris. In 1906, the company issued a catalogue containing 17 machines, both front and back-mounted types, priced from 15 to 222 FF. Also listed were several coin-ops and an internal-horn model. Although the catalogue claimed "manufacture française," the machines were unmistakably German in appearance: some were clearly identifiable as Adlers, and one was the spit and image of a Hymnophone.

The French talking machine industry experienced upheaval during the First World War. When peace returned, the cylinder phonograph was no longer viable in France. Pathé and all other French manufacturers had made instruments which played only the two-minute cylinders that were already obsolete when the war began. Although four-minute cylinders had been available the United States since 1908, no appreciable market arose in France for the improved four-minute cylinders or machines to play them. Before the end of the 1920s, the sapphire-ball disc, too, would become extinct. Pathé,

3-8. The Théatrophone betrays its provenance in the styling of its pretty decal. Ostensibly an imitation of Graphophone decoration, the exact execution of the decal suggests Swiss origin. The style of the motor and cabinet reinforce this impression. Note the added gearing on the right which allowed the machine to be wound with a crank rather than a key. *Courtesy of Jean-Paul Agnard.* (Value code: G)

seeking to remain competitive in a world where lateral cut was the industry standard, abandoned the vertical system. It began manufacturing lateral cut discs, first in the United States, then in France.

Today, we are left with ample evidence of the industriousness of British and European talking machine companies. Despite the depredations of war, climate and neglect, the world is still rich with their products. It is the legacy of these firms to lend color, charm, character and non-conformity to the antique phonograph collecting field.

3-9. The German Excelsior company offered a wide variety of talking machines during the first decade of the twentieth century. This coin-operated table model machine, with curved glass top to display the mechanism, was designed to play either "regular" sized or 5" diameter cylinders. Note the articulated metal tube which connects the reproducer with the horn. *Courtesy of the Domenic DiBernardo collection.* (Value code: D)

3-10. This charming European clock gives little suggestion of its true nature, except for the large winding key that protrudes from the side of the wooden case, a rather unusual place for it to be. *Courtesy of Jean-Paul Agnard.* (Value code: VR)

3-11. Hinging back the cornice of the clock reveals that the large key is for winding the mainspring of a custom cylinder phonograph mechanism. The separate clock mechanism is wound through the face by the smaller key, as is customary. The phonograph acts as an alarm, playing a two minute cylinder at the selected time. Talking or musical clocks were among the first marvels envisioned after the advent of recorded sound. The easiest application, of course, was as seen here: substitution of music for a bell or chime. *Courtesy of Jean-Paul Agnard.* (Value code: VR)

3-12. French-constructed "Puck" cylinder phonographs were characterized, in part, by their silver-painted metal bases. This particular French version is further distinguished by a raised medallion depicting a lion in profile. Three types of "Lion Pucks" are known: facing right, facing left and facing forward. "Pucks" can generally be found equipped with one of three types of horn: straight zinc (shown), japanned steel flower and spun aluminum. Most European distributors offered a choice of horns. *Courtesy of Jean-Paul Agnard.* (Value code: G)

3-13. This "La Fauvette" (the warbler) machine from France, circa 1904, featured a complex and unusual upper casting. Mechanisms like this were expensive to manufacture, and showed more commitment to the product than did the endless European copies of the Graphophone "Eagle." Although the general layout of the machine suggested an Edison "Standard," a certain Graphophone influence can be seen in the reproducer carriage. Perhaps the most peculiar feature of this machine is the position of the feedscrew at the front of the main casting. A number of interesting machines were sold under the "La Fauvette" name, as well as a line of two-minute cylinders. *Courtesy of Jean-Paul Agnard.* (Value code: F)

3-14. An utterly charming French adaptation of an Edison "Gem" Model "A" Phonograph, which leaves one wondering, "Why?" A great effort has been made to conceal the black cast-iron body of the original machine from view. The over-all wooden cover with which the "Gem" was originally sold has been cut roughly in half, horizontally. To the bottom half of this bisection, additional molding has been appended to hide the radical surgery. However, the presence of the Edison banner decal prevented disguising the obviously truncated lid. A crank, instead of the usual winding key, completed the transition of a modest machine into a slightly elevated incarnation. A lot of work for a little improvement in appearance, but we thank talking machine and piano dealer Léon Auffray of 39 rue Saint-Aubin, Angers, for giving us such a delightful peculiarity. *Courtesy of Jean-Paul Agnard.* (Value code: G)

3-15. In 1898, Columbia introduced a cylinder record (the "Grand") which was of normal 4" length, but had a diameter dramatically enlarged to 5". The theory behind this development was increased surface speed, not increased playing time. When played at the same number of revolutions per minute as a standard 2 1/8" diameter cylinder, a cylinder of larger diameter would cause the recorded information to pass more quickly beneath the reproducing stylus. This resulted in better "fidelity," which was coupled with improved depth of groove for greater volume. Although many companies quickly released their own versions of the 5" cylinder record (Edison's was the "Concert"; Pathé had the "Stentor"), the concept of the very large diameter cylinder made the greatest impact on the Pathé company of France. Perhaps the most profound evidence of this can be seen in the "Le Céleste." *Courtesy of Jean-Paul Agnard.* (Value code: VR)

3-16. The "Le Céleste" employed the over-all largest cylinder record ever put before the public, measuring 5" in diameter by 9" long. In its studio, Pathé used even larger diameter wax cylinders as master records. The company also originated and heavily promoted a 3 1/2" diameter "Inter" ("Intermédiaire") cylinder in Europe and Great Britain. The "Le Céleste," however, was Pathé's *grande dame.* Two versions were sold. The first, appearing in the 1900 catalogue with a glass horn, was the larger of the two. The second version (shown) appeared in the 1903 catalogue, and was the more ornately decorated. It is likely that this particular machine dates from a year or so later, since it incorporated the built-in "floating horn" system which Pathé did not yet illustrate in its 1903 catalogue. *Courtesy of Jean-Paul Agnard.* (Value code: VR)

3-17. Stollwerck was a Belgian confectioner that hit upon a clever idea for promoting its sweets. The company took a cue from the "Toy" Graphophone of 1899-1900 that had played small diameter, vertically recorded wax discs. If discs could be pressed of wax (actually wax-like soap), then why not chocolate? Chocolate had already been molded into every other effigy, from money to farm animals, so edible records were not far-fetched. Around 1903, Stollwerck marketed a simple lithographed disc talking machine run by a clockwork motor, which played 3 1/8" diameter vertical cut chocolate records. A reproducer identical to the one used on a "Puck" cylinder phonograph directed the sound into a brightly decorated conical horn. Some have speculated that the recording was actually pressed into a foil covering which originally may have enwrapped each disc. Most extant discs show no signs of such a covering, though it must be noted that wax or composition discs were soon introduced to be used on the machines. A few years later, the same device reappeared in the French market under the name "Eureka." (Value code: VR)

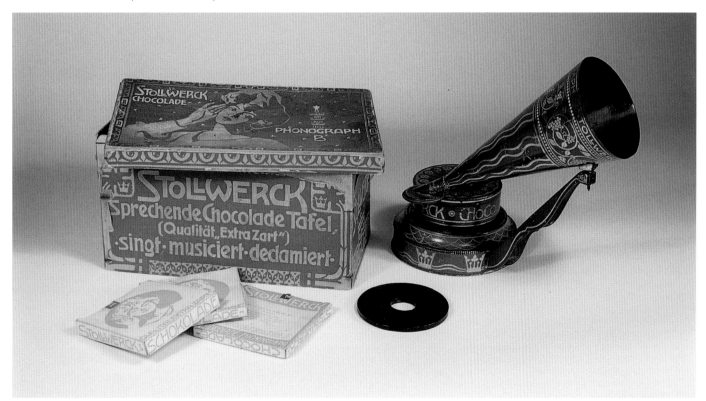

3-18. A box of discs to fit the "Eureka" talking machine, by which the mechanism most commonly associated with Stollwerck was known in France. These discs were "wax," vertically recorded. *Courtesy of the Domenic DiBernardo collection.*

3-19. The Belgian *chocolatier* Stollwerck was not really in the phonograph business, except to distribute the little clockwork disc machines sold under its name. Illustrated here is a turn-of-the-century cabinet from which a shopkeeper dispensed Stollwerck confections.

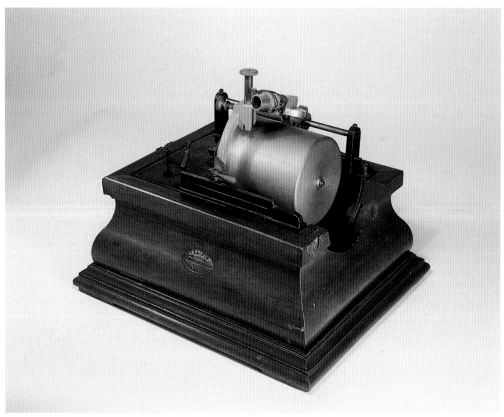

3-20. Cylinder talking machine mechanisms in exaggerated mahogany cabinets marked "Verifié (verified), Paris" were the joint work of the enormously productive Excelsior company of Germany, and Paris distributors Laurant & Salomon. This 5" diameter cylinder machine, circa 1904, borrowed both Edison and Columbia technical elements. The most striking features, however, were the design of the recessed mandrel (Excelsior), and graceful hardwood cabinetry (Laurant & Salomon). Excelsior mechanisms were sold under a considerable number of brands throughout Europe and Great Britain. *Courtesy of Ray Phillips.* (Value code: E)

3-21. Another "Verifié" machine in an absolutely luscious mahogany cabinet. The mechanism itself was an Excelsior model similar to a Graphophone Type "Q." Incredibly, this diminutive music maker was set atop a largely empty cabinet! An accessory drawer entered the base cabinet from the right side, but surely the reason for the elephantine scale was to lend visual punch. Like many talking machine companies through the years, Laurant & Salomon found that wood, no matter how beautifully finished, was still cheaper than metal. In this instance, the company imported the little Excelsiors and inserted them into domestically built cabinets. The unusual lid clasps were stamped with a French patent designation. (Value code: G)

3-22. Never was there a phonograph for the masses like the uncomplicated "Puck." How many hundreds of thousands—millions—were produced in Europe between about 1900 and 1914 is a question beyond speculation. Many variations, both slight and significant, are known. Here, the lyre-shaped base, most commonly cast in iron, has been stamped out of steel. On the base can be seen a rectangular "licensing notice" which was affixed to the "Pucks" imported into the United States and sold through or under license from the Columbia Phonograph Company. The horn displays an unusual *bombé* petal design. *Courtesy of Jean-Paul Agnard.* (Value code: H)

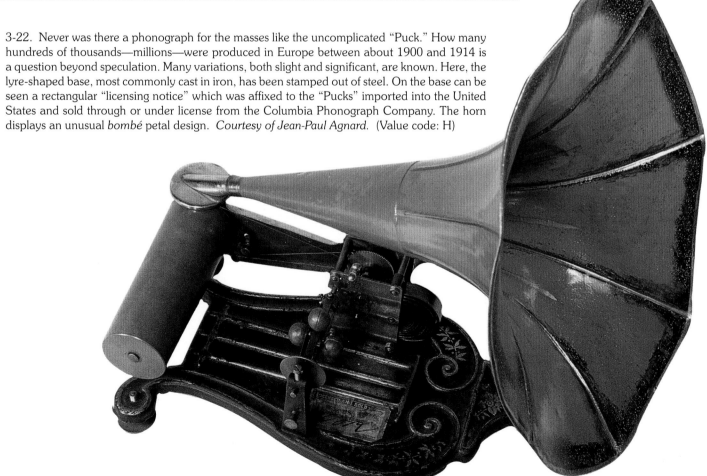

3-23. A printer's cut of a Puck phonograph, used in newspaper advertising circa 1905, measuring 1 7/8" x 2 1/8". *Courtesy of Mr. R. Chase.*

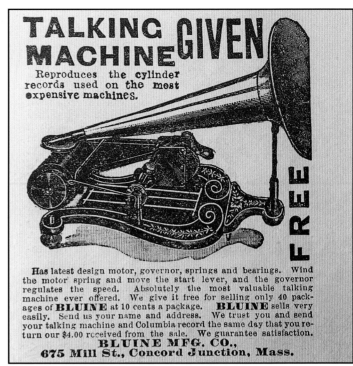

TALKING MACHINE GIVEN

Reproduces the cylinder records used on the most expensive machines.

Has latest design motor, governor, springs and bearings. Wind the motor spring and move the start lever, and the governor regulates the speed. Absolutely the most valuable talking machine ever offered. We give it free for selling only 40 packages of **BLUINE** at 10 cents a package. **BLUINE** sells very easily. Send us your name and address. We trust you and send your talking machine and Columbia record the same day that you return our $4.00 received from the sale. We guarantee satisfaction.

BLUINE MFG. CO.,
675 Mill St., Concord Junction, Mass.

3-24. An American advertisement which offered a "Puck" (and one Columbia record) free for selling 40 packages of "Bluine."

3-25. The "Sylvia C" was a Swiss import marketed in Great Britain. Note the over-all nickel plating favored in the United States and Switzerland. The novel manner in which a mechanism originally designed to be wound from the side by a key has been adapted to use a vertical winding handle suggests Excelsior's "La Favorita," also marketed in Britain, which employed a similar cranking system. *Courtesy of Jean-Paul Agnard.* (Value code: G)

3-26. In Germany, variations on the simple theme of the "Puck" cylinder phonograph seem to have defied all limits of imagination. Here we see the usual puny playing mechanism encased in a wooden box, with carrying handle to facilitate transportation. A hatch opened to allow the record to be played, using a typical "Puck"-type reproducer and straight zinc horn. *Courtesy of Jean-Paul Agnard.* (Value code: G)

3-27. Typical of the clever and compact engineering of German cylinder talking machines is this Adler model, circa 1903. Note the chain drive, which Adler used on a number of its machines, though it was not tried elsewhere. Frequently, small instruments such as this bore no brand name. Sometimes they sported decals which might proclaim, "Lucca," "Dulcetto" or simply "Phonograph." *Courtesy of Jean-Paul Agnard.* (Value code: H)

3-28. The German Koh-I-Noor phonograph reflected the distilling of technology that produced swarms of miniature mechanisms in brightly-marked cabinets. *Courtesy of Jean-Paul Agnard.* (Value code: H)

3-29. In 1905, the Edison Bell company of Great Britain began manufacturing its "Gem" Phonograph as a companion to its popular "Standard" model. Although the "Gem" and "Standard" names were shared with American-made Edison machines, the instruments themselves were really rather different. To understand how a litigious fellow like Mr. Edison could have been brought to the usurpation of his cherished trade names, it is necessary to examine a little history. In 1903, a ten-year period ended during which Edison Bell had ruled the British talking machine market through control of all fundamental Edison and Bell-Tainter patents. The Edison company in the United States had come to rue the day it allowed Edison Bell such a powerful license, and in 1903 Edison's National Phonograph Company set up a branch in England. What had been sizzling resentment flared into open warfare, and over the next few years National Phonograph tried in every way to discourage Edison Bell from drawing any associations with American products, or especially trade names. Eventually, Edison Bell dropped the "Standard," "Gem" and "Homestead" line they were manufacturing, and substituted cylinder machines with names which were not offending to the Edison company. The old feud, however, was never forgotten, especially since Edison Bell seemed to enjoy thumbing its nose at the competition. *Courtesy of Jean-Paul Agnard.* (Value code: H)

3-30. An Edison Bell catalogue, circa 1905. The nationalistic sloganeering in which the company frequently engaged is here supported by the fact that the advertised machines were produced in England. *Courtesy of Sam Sheena.*

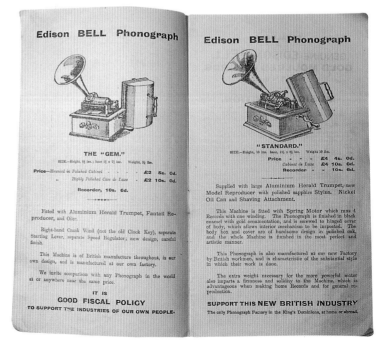

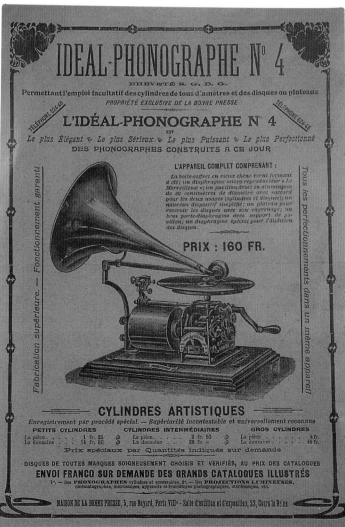

3-31. Maison de La Bonne Presse was a Paris religious publishing house for which talking machines became a significant sideline. It was not uncommon for publishers, especially of magazines and newspapers, to offer an up-to-date novelty such as the phonograph to entice subscribers. However, Bonne Presse expanded its phonographic activities to the level of becoming a major presence in the French talking machine and record market. The "Idéal-Phonographe No.4" illustrated here played either cylinder or disc records. It was a more complicated variation of the company's popular "Idéal" model that played cylinder records exclusively, either 5" diameter or standard-sized. *Courtesy of Jean-Paul Agnard.* (Value code: VR)

3-32. A 1905 advertisement for the combination "Idéal" touts the machine's advantages. The fact is, the "No.4" offered the customer a perfect way to navigate the choppy waters of the European talking machine market. Firstly, the machine could play 5" diameter cylinders, which persisted in Europe longer than in the United States. Furthermore, by removing the large mandrel standard-sized cylinders could be used. If the customer were so inclined, a Pathé "Inter" mandrel could be slipped on to play Pathé's 3 1/2" diameter cylinders. Lastly, the machine was capable of playing any lateral cut disc record. Since the vertical cut disc had not yet been popularized in Europe, the "Idéal No. 4" was an instrument true to its name: *ideally* suited to reproducing any record. *Courtesy of Jean-Paul Agnard.*

3-33. Illustrated here is a machine very reminiscent of the "Biophone," a metal-encased variation of the "Puck," made in Germany for sale in Britain. This view shows an example whose metal cabinet renders the cylinder nearly inaccessible! The cylinder must be played through a slot, with a greatly elongated stylus. *Courtesy of Jean-Paul Agnard.* (Value code: G)

3-34. The German-made "Puck" instruments were offered with a variety of horns and base designs. This example, with another version of the "lion" base, displays a delicately-shaded horn of unusual color. *Courtesy of Jean-Paul Agnard.* (Value code: G)

3-35. Throughout the centuries, the sinuous female form has inspired designers, and they have insinuated its image practically everywhere. No exception was the humble Puck phonograph. Two "woman-inspired" versions existed. The one shown here was known as the "Nymphe." Though rightly a mermaid, she was the creation of Georges Carette and Company of Nurenburg, Germany. Indeed, she might have stepped out of a Mannerist painting, such as Agnolo Bronzino's sixteenth-century *An Allegory of Venus, Cupid, Time and Folly*. The Carette catalogue offered a choice of three horns for this machine: straight zinc, spun aluminum, or flower. In Great Britain, "No.856S," with black japanned base, sold for 6/- (shillings). For another six pence the customer could have "No.856G": "handsomely gilt" like the one here. (Value code: G)

3-36. A close-up showing the Carette attribution in relief.

3-37. It is practically incomprehensible how Excelsior of Germany could have produced so many different models of cylinder talking machine. Most were virtually indistinguishable, except for the brand name under which they were sold. However, this interesting "New Century" machine, intended for the British market, deviated from other Excelsior products in one splendid aspect: the governor exited the motor housing like a rocket taking flight. *Courtesy of Jean-Paul Agnard.* (Value code: G)

3-38. In a French cylinder talking machine market strongly dominated by Pathé, a handful of other French brands competed with surprising vitality. The "Phénix" (phoenix) brand not only vied with Pathé, it promoted its own specially-sized cylinders, approximately 3 3/8" in diameter. A prominent feature of many "Phénix" instruments was a circular dial with speed control calibrations, seen here at the left top of the motor housing. This particular example is housed in a "reversible" cabinet, although a similar mechanism was sold fixed to a base board, and covered by a rectangular lid. The "Phénix" appeared in both "floating horn" design (shown) and traditional "Graphophone" style, with a horn directly mounted on the reproducer carriage. *Courtesy of Michael Sims.* (Value code: G)

3-39. An odd cousin of the "Phénix" was the French "Excelsior" (not to be confused with Excelsior of Germany) sold by G. Maleville of Libourne (sole European sales agent). The left-hand celluloid identification plate announced, "…seul phonographe à voix naturelle" (the only natural voice phonograph). The visual incarnations of cylinder and disc talking machines were distinct until the mutual adoption of "back brackets" supporting flower horns began to merge their respective images, around 1907. This "Excelsior," circa 1904, was a clue to the new direction, borrowing heavily from the earlier "front-mounted" disc styling. It played the special "Phénix" brand cylinders. (Value code: G)

3-40. The basic arrangement of the Puck phonograph, including lyre-shaped base, was actually patented by Gianni Bettini (No.618,390, January 31, 1899), who, throughout his career in the talking machine industry, seemed to experience terrible luck in holding onto his clever ideas. The Puck in its simplest form may have been *reductio ad absurdum*, but it was just enough of a phonograph to satisfy thousands who could afford no better. Of course, manufacturers soon set to work "dressing up" the Puck and attempting to conceal its identity. Here, a four-footed version, suggestive of the decoratively cast base occasionally sold with the Columbia Type "Q" Graphophone, looks as if it might be capable of taking a stroll. (Value code: G)

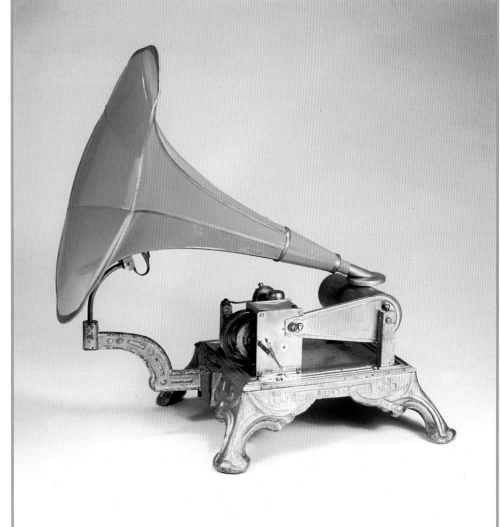

3-41. Considering the ubiquitous nature of the Puck phonograph, it seems inevitable (though somewhat absurd) that it would have been adapted to coin operation. Here we see an example of one such rare bird: really more of a Dodo than a lark. The glazed, walnut cabinet featured the world's smallest signboard. Note the traditional Puck key has been replaced by a crank, albeit a tiny one. Cabinet measures 8 7/8" wide, 15 7/8" deep and 10 5/8" high, not including signboard. (Value code: VR)

3-42. The interior mechanism of the coin-op Puck, though mounted on a casting of abnormally large dimensions, still maintained all the recognizable Puck components. The worm feed-gear was, of course, a necessity of the coin function. The "tone arm" was a rarely seen feature used on certain elaborately designed Pucks such as the "Lorelei." Because of the Puck's customary "backward" orientation, the machine played (ordinary two-minute cylinders) from right to left. This machine, made in Germany, was adapted to an English penny. Although we may marvel at the mind which modified a marginal mechanism for commercial use, we are left with further evidence of the Puck's universality.

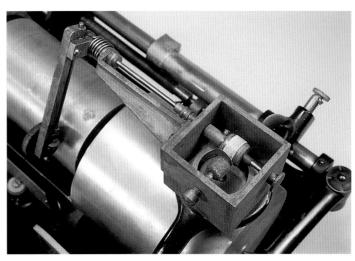

3-43. Like Daniel Higham, Edison experimented with the amplification of diaphragm vibrations by means of friction. Whereas Higham used amber to create a high degree of resistance, Edison employed chalk in his Electromotograph apparatus. This device was intended to amplify the sound of a telephone receiver. Here we see the Electromotograph adapted for recording telephone transmissions on an Edison electric 5" cylinder Phonograph (known as the "Opera"). The amplifying unit is in the box located above the recording head. *Courtesy of the Edison National Historic Site.* (Value code: VR)

3-44. The Electromotograph consisted of a rotating chalk wheel and a charged metal electrode that was connected to a diaphragm. An adjustable rubber pressure pad kept the electrode in contact with the chalk. The physical reaction resulting from this friction amplified a telephone transmission in order to make it loud enough to record. The original inspiration was to facilitate creation of a recordable telephone communications network for railroads. When Daniel Higham's friction loud-speaking system for talking machines came into use, Edison sued on the basis of the Electromotograph. Edison, whose system was really quite different than Higham's, and far less effective, lost in court. Ironically, when Edison prepared to introduce his Kinetophone talking movies (exhibited 1913-1916) he adopted Higham's proven method of amplification. *Courtesy of the Edison National Historic Site.* (Value code: VR)

3-45. The Victor Talking Machine Company's "Premium" series was a fascinating and unpredictable melting pot for whatever odds-and-ends the firm wanted to dispose of. Certainly, the rarest and most peculiar of this strange breed was the Victor "P No.3" (serial number 766 shown). This instrument should not be confused with the Victor "P3," a machine in a squarish oak cabinet that employed a very different cabinet and motor. The "P No.3," circa 1905, had an inexpensive, low wooden cabinet, stained mahogany. The hodge-podge motor was held by an oxidized rectangle of sheet metal that formed the top of the cabinet. The horn support arm was oxidized, and the considerably offset turntable measured 7" in diameter. In all, the machine appeared to have been pieced together rather obviously, which of course it was. *Courtesy of Walter and Carol Myers.* (Value code: E)

3-46. One of several recording machines remaining at Edison's factory, all of which were adapted from home entertainment models. Here, an Edison "Triumph" Model "A" has been fitted with a lead-sheathed flywheel, a magnifier with electric light and a surprising carriage. At first glance, it appears that this Edison has been fitted with Graphophone parts. The recording head mimics the "floating" arrangement of a Graphophone, and the horn connection suggests the same influence. The proprietary design of Edison recording heads sold to the public seems to have been inadequate for Edison's own factory use. *Courtesy of the Edison National Historic Site.* (Value code: VR)

3-47. Three of a previously unseen cache of Edison factory recording heads. Each has been specially adapted for a particular application. It seems ironic that Edison was employing these clearly Graphophone-like devices in his recording lab, considering his antipathy to the Graphophone interest. *Courtesy of the Edison National Historic Site.* (Value code: VR)

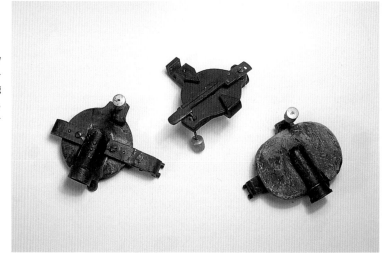

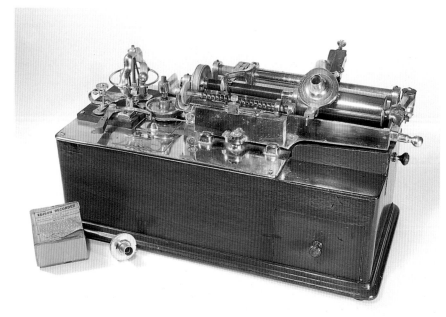

3-48. A gold-plated Edison "Balmoral" Phonograph, circa 1905, in a mahogany cabinet. This machine (No. 39404) was further distinguished by a Model "D" repeating attachment and a matching gold-plated recorder. The "Balmoral" cost $65.00, and the gold-plating was a $50.00 option. The mahogany cabinet added another $4.00, raising the price of this example to a princely $119.00. *Courtesy of Ray Phillips.* (Value code: VR)

3-49. This solid gold cylinder record on a sterling silver platform was presented to Thomas Edison during the July 17-20, 1906 convention of talking machine jobbers (distributors) hosted by the National Phonograph Company. The cherubs seated on the platform represent music, art and progress. The inscription to the right reads: "PRESENTED TO THOMAS A. EDISON, JULY 18, 1906 BY THE EDISON PHONOGRAPH JOBBERS OF THE UNITED STATES AND CANADA AS AN EXPRESSION OF THEIR PERSONAL ESTEEM." To the left is a bas-relief depiction of the famous Massani painting of the old couple listening to "The Phonograph." During the presentation, the gold record was played:

Mr. Edison. The record of pure gold which addresses these words to you is a gift from the jobbers engaged in the distribution of the Edison Phonograph and Records. They have come from the East, the West, the North and the South of this fair land of ours. They are your loyal and abiding friends, your ambassadors of commerce, whose mission it is to distribute your product to the four quarters of the globe. By the touch of your colossal inventive genius you have created industries giving employment to countless thousands the world over. The wheels of commerce occupied in the production of your inventions sing a never-ending song of praise to your magnificent achievements. To the seven existing wonders of the world you added the Phonograph, which is the eighth wonder of the world. It speaks every language uttered by the human tongue, and in the field of language study it is the greatest educator the world has ever known. With song and story it will continue till the end of time to entertain the multitudes of the earth who place the name of Thomas A. Edison at the head of the column of the world's greatest captains of industry. [Band rendition of Auld Lang Syne.]

Overall height of the gold cylinder and silver platform is 8 3/4". A turned ebony base was originally part of this presentation, elevating the entire affair to approximately 12" in height. Later during the convention, each attendee received a replica cylinder, molded in conventional black "wax." *Courtesy of the Edison National Historic Site.* (Value code: VR)

3-50. In 1906, when Pathé entered the French disc talking machine market, the company drew its inspiration from the cylinder technology it had perfected over the better part of a decade. Direct resemblance to cylinder models could be seen in Pathé's earliest disc machines: they incorporated ebonite "floating" reproducer heads, spun aluminum horns and walnut cabinets decorated with a cylinder talking machine, a rooster and the motto "Je Chante Haut et Clair" (I sing loud and clear). Although Pathé soon developed a metal soundbox and other design features which reflected conventional disc talking machines, this 1907 Model "A" continued to betray cylinder kinship in its decal and its flower horn, which was identical to ones sometimes sold on Pathé cylinder phonographs in Great Britain. *Courtesy of Michael and Suzanne Raisman.* (Value code: F)

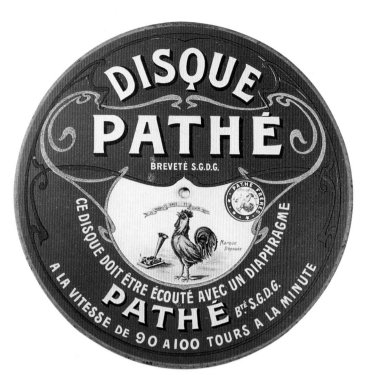

3-51. The earliest Pathé disc records had a fragile wax recorded surface laminated to a sturdy backing of composed material resembling concrete. Because the vertically recorded wax playing surface could not withstand the significant pressure exerted by conventional disc soundboxes, the very first Pathé disc reproducers were direct adaptations of lightweight cylinder machine heads. Pathé discs were single-faced, like disc records the world-over in 1906. Pathé, however, was unique in adding colorful imagery to the non-playing side of its records.

3-52. A nickel-plated Edison "Triumph" Model "B" (introduced 1905), in a custom mahogany cabinet with inlay. *Courtesy of John Woodward.* (Value code: VR)

3-53. The Edison "Windsor" Phonograph was a coin-operated machine based upon the venerable Class "M" mechanism. First offered in 1904, this example featured the rounded corner posts introduced around 1906. *Courtesy of Jean-Paul Agnard.* (Value code: A)

3-54. The Edison "Windsor" Phonograph in its opened position. Storage batteries were kept in the lower compartment. Note also the coin chute and locking coin box. The "Windsor" originally sold for $65.00. *Courtesy of Jean-Paul Agnard.*

3-55. The coin-operated Edison "Acme" Phonograph was introduced in April 1906 as the first machine electrically powered by AC offered by Edison. The motor operated on 104-110 volt 60 cycle alternating current. It was identical in appearance to the "Eclipse" Phonograph, which used a 125-volt DC motor. Both the "Acme" and the "Eclipse" originally sold for $65.00.

ALL FOR A PENNY!

DRAWN BY WALTER APPLETON CLARK

3-56. A magazine illustration from 1906 entitled "All For A Penny." From the 1890s through the turn of the century, coin-operated talking machines were uncommon enough to command a nickel. By 1906, the proliferation of coin-slot music machines in public places and arcades had driven the price down to a penny.

3-57. An Edison dealer window decal dating from the 1905-1907 period. *Courtesy of the Charles Hummel collections.*

3-58. Here is a machine that has never been documented: a previously unknown version of the Columbia Disc Graphophone Type "AU." Columbia introduced the "AU" in 1904. At only $12.00, the little "open-works" machine was an instant success. Thousands and thousands were built both as Columbias and client brands. In addition, the small but effective motor was adapted to power a number of client machines with wooden cabinets (such as the Standard "X2"). The "AU" always carried the identical equipment: 7" diameter turntable, aluminum "traveling arm," brass horn elbow, Columbia "Analyzing" soundbox and 16" black japanned horn—until now. The machine illustrated here employed no aluminum traveling arm. Instead, a hinged support was affixed to the body of the horn, and the soundbox pushed directly into the customary Columbia brass elbow. This basic horn arrangement had been used by Columbia previously. One of the earliest versions of the Disc Graphophone Type "AK" had the same direct-to-horn pivot. However, the "AK" elbow (which was aluminum) and the soundbox (which was an earlier style) were quite different than those employed here. It can be surmised that Columbia, which was extremely imaginative when it came to trimming expenses, saw an opportunity in this unusual "AU" to eliminate the cost of the aluminum traveling arm. (Value code: G)

3-59. Are there "Doubting Thomases" among you? Please observe the two soundboxes illustrated here. The one at the bottom is a conventional Columbia "Analyzing" soundbox. The one at the top is from the uncatalogued "AU." The location of the positioning pin always found on the reverse side of a Columbia "Analyzing" soundbox differs between these two examples. The customary purpose of this pin was to secure the soundbox to the aluminum traveling arm. Since the newly discovered version of the "AU" did not employ an aluminum arm, the factory changed the location of the pin to allow it to ride against the ridged seam of the brass elbow, and thereby hold the soundbox in correct playing position. The soundbox tube or "throat" was also reduced in length, since it did not pass through the thickness of an aluminum arm. Why Columbia failed to take widespread advantage of the "simplified AU" is anybody's guess.

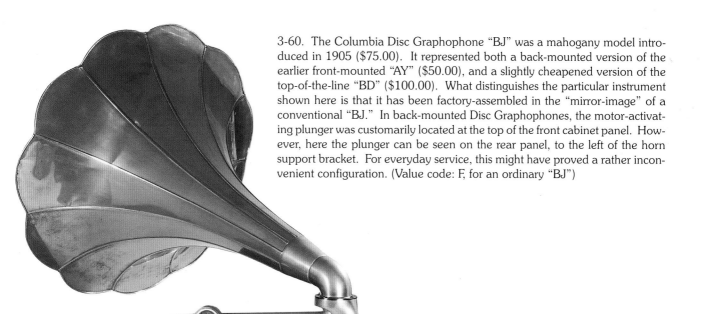

3-60. The Columbia Disc Graphophone "BJ" was a mahogany model introduced in 1905 ($75.00). It represented both a back-mounted version of the earlier front-mounted "AY" ($50.00), and a slightly cheapened version of the top-of-the-line "BD" ($100.00). What distinguishes the particular instrument shown here is that it has been factory-assembled in the "mirror-image" of a conventional "BJ." In back-mounted Disc Graphophones, the motor-activating plunger was customarily located at the top of the front cabinet panel. However, here the plunger can be seen on the rear panel, to the left of the horn support bracket. For everyday service, this might have proved a rather inconvenient configuration. (Value code: F, for an ordinary "BJ")

3-61. Because the cabinet has been re-oriented by 180 degrees, the winding crank of this "BJ" is located on the opposite (left) side of where it would customarily be found. Needless to say, winding the motor might have been a bit awkward for someone used to "normal" talking machine orientation. The two Columbia decals seen here were customarily applied to the "BJ." There are no signs of any untoward modification of this machine, and a paper tag identifying the machine as a "BJ" is affixed to the underside of the cabinet. Perhaps this is an early example of the "BJ," with less thought having been directed toward the orientation of the components, since the direction of the horn was less a factor in the front-mounted machines from which the "BJ" immediately evolved. Perhaps, again, this machine was specifically commissioned by an exhibitor, for whom it would have been most convenient to stand *behind* the machine, facing an audience. Viewed within this context, the positioning of the components suddenly makes perfect sense.

3-62. In 1900, the sales and distribution of Emile Berliner's Gramophones were thrown into sudden turmoil. Berliner's primary distributor, Frank Seaman, after many months of dissatisfaction with the policies of the Berliner Gramophone Company, suddenly allied himself with the rival American Graphophone Company. Under the protection of Graphophone patents, Seaman began offering the Zonophone, a competing instrument (in which he had unsuccessfully attempted to interest the Berliner Company). Emile Berliner found himself in the lamentable position of no longer having a sales network. After a brief effort at Gramophone retailing in the Philadelphia area, Berliner moved his operations to Montreal, Canada. There he would remain and flourish under a cross-licensing agreement with the Victor Talking Machine Company. This Canadian Berliner Model "A," circa 1901, is the veritable spit and image of the "Improved" Gramophone of 1897. *Courtesy of William G. Meyer.* (Value code: E)

3-63. The Canadian Berliner "D" is a peculiar amalgam of Gramophone historical design features. The later style wooden traveling arm, the plunger brake and open-faced soundbox are elements of the period 1902-1905. The black japanned horn with double stripe, leather elbow, and early motor with a vertical crank all suggest the late 1890s. *Courtesy of the Domenic DiBernardo collection.* (Value code: F)

3-64. The Canadian Berliner "C," circa 1903, was the largest and best-appointed of the company's "front-mounted" machines. It used a 10" turntable and was available in either oak or mahogany (shown). Of note are the uniquely-shaped escutcheons for the crank and speed control, characteristically found on Canadian Berliner machines. Berliner equipped this machine with an advanced soundbox of large diameter, but seldom abandoned the primitive leather elbow he began using in the early 1890s. *Courtesy of the Domenic DiBernardo collection.* (Value code: F)

3-65. Outwardly, this Canadian Berliner "E" appears to be a Victor "Royal," but several features distinguish it. Although some of the metal parts have an oxidized bronze finish, the cabinet lacks the metal corner plates and plain molding of the Victor "R." The metal top of the cabinet is stamped steel, whereas Victor used cast iron. Furthermore, the plunger brake was an improvement which Victor never adapted to the "Royal." Note the characteristic narrowing of the wooden arm near the soundbox, which is found only on Canadian machines. *Courtesy of the Domenic DiBernardo collection.* (Value code: G)

3-66. The Canadian Berliner "F" was an inexpensively-constructed machine which reduced essential elements to a minimum. The wooden traveling arm was completely eliminated, necessitating the rare use in Canada of a metal elbow to hold the "Automatic Grand" soundbox. *Courtesy of the Domenic DiBernardo collection.* (Value code: F)

3-67. The appearance of the Canadian Berliner "J," circa 1909, was improved by wood appliqués such as had been used by the Talkophone Company in the United States. It was the largest of the Canadian Berliner company's "back-mounted" machines, and is here equipped with a wood-grained Spaulding Linen Fibre Horn. *Courtesy of the Domenic DiBernardo collection.* (Value code: F)

3-68. A close-up of the wood-grained horn of the Canadian Berliner "J." This was manufactured by J. S. Spaulding & Sons Company of Rochester, New Hampshire, and retailed for $8.00 (US). The *Talking Machine World* of February 1908 included a Spaulding advertisement stating that the company's Canadian distributor was the Berliner Gramophone Company of Montreal. *Courtesy of the Domenic DiBernardo collection.*

3-69. The Canadian Berliner "K" bore strong similarities to the Victor "I" except for the horn and tone arm. The horn, with its pointed profile and vivid red color was in sharp contrast to the drab Victor flower horns. The cantilevered tone arm presented an interesting alternative to the conventional Victor U-joint, for placing the soundbox at rest. Like a similar tone arm arrangement first used by Hawthorne & Sheble, the loose-fitting joint in the arm meant an immediate loss of sound quality. *Courtesy of the Domenic DiBernardo collection.* (Value code: F)

3-70. A close-up of the "K" horn shows that the decal is virtually identical to a Victor decal, except for the company name. *Courtesy of the Domenic DiBernardo collection.*

3-71. The Canadian Berliner "G" looked like a dressed-up version of the Victor "E" (the Canadian "B" was a dead ringer). It is interesting to note the use of an oxidized support arm and elaborate base molding, features that were never combined on Victor machines. *Courtesy of the Domenic DiBernardo collection.* (Value code: G)

3-72. If we proceed from the assumption that the initials in the name of the Canadian Berliner "GT" stood for "Type G, Tone Arm," the logic of this can be seen in the photo. The cabinet proper much resembled the front-mounted "G," except for the strikingly stepped base molding. Coming as it did toward the end of the Canadian company's period of semi-autonomous designs (1909), this machine more closely conformed to ordinary Victor construction. This trend would culminate in the last of the external-horn Canadian Berliners (1910-1912). These later machines were undisguised Victor "I"s, "II"s, and "III"s (with slight variations of back bracket, tone arm, elbow) identified by a little plate listing the seller as the Berliner Gramophone Company. *Courtesy of the Domenic DiBernardo collection.* (Value code: G)

3-73. The Douglas Phonograph Company of 89 Chambers Street, New York City, was an aggressive merchandiser of talking machine products from 1900 through 1907. One of its specialties was custom cabinetry for Victors. Shown is a Victor "V" (No.22,186) housed in a Douglas cabinet. This design was illustrated in the January 1907 issue of *Talking Machine World*. Douglas is best remembered for incorporating external-horn talking machine mechanisms into the tops of floor-standing record storage cabinets. Smaller custom cabinets such as this one are less frequently encountered. *Courtesy of Roger Bodenheimer.* (Value code: VR)

3-74. The Mills Novelty Company of Chicago, Illinois, manufactured this coin-operated talking machine with fortune attachment circa 1907. Although the particular machine shown accepted nickels, most coin-operated talking machines by this time operated for a penny. Mills enticed customers to part with their pocket change by adding a gimmick to the commonplace musical entertainment; the machine also told fortunes. Some arcade devices gave out printed cards with each audition, entailing extra cost which this particular model cleverly avoided (see next illustration). *Courtesy of the Sanfilippo collection.* (Value code: A)

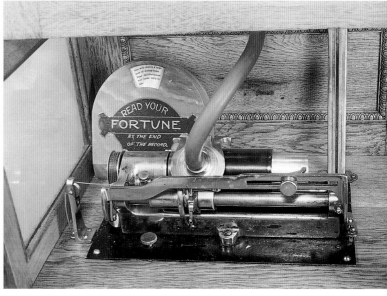

3-75. An interior view of the Mills musical fortune-telling machine. The playing mechanism is a Type "BG" Graphophone, capable of playing 6" long Columbia "20th Century" cylinders. Rather than dispensing printed fortunes which required replenishment, this example featured a rather simple attachment whereby a new fortune was displayed each time the machine was played. The string at the left was pulled by the reproducer carriage as the machine played. When the carriage reached the end of the record, a weight was tripped that advanced the wheel to the next fortune, of which there were nine. *Courtesy of the Sanfilippo collection.*

3-76. The Mills Novelty Company of Chicago produced this "Illustrated Song Machine" in 1907. It incorporated in its towering frame a Columbia "BG" Graphophone cylinder mechanism (just visible through the window at the bottom) which provided musical accompaniment for a series of stereo views. In this instance, the novelty of the mechanism was sufficient to command the price of a nickel. Note the "retractable" listening tubes on either side of the eyepiece. *Courtesy of the collection of Howard Hazelcorn.* (Value code: VR)

3-77. The Edison Bell "Elf" was one of the 1907 series of British-made machines which continued to suggest traces of American Edison design, even after a considerable period of estrangement. *Courtesy of Jean-Paul Agnard.* (Value code: H)

3-78. The Excelsior company of Germany produced a series of cylinder phonographs specifically aimed at the British market, during the post-1905 period. "La Favorita" (named for Donizetti's opera) was one of them. The influence of Columbia's Type "Q" Graphophone was apparent in this low-end model. However, the vertical winding arrangement and "reversible" cabinetry lent an unmistakably European air. *Courtesy of Jean-Paul Agnard.* (Value code: H)

3-79. This machine is representative of the many, many varieties of disc talking machine which were produced in Germany during the first decade of the twentieth century. The cabinet is large to give the machine substance, though the motor is relatively small. The free-standing columns reflect a decorative element found in upper-price machines. The brass appliqués were cheap to manufacture, yet lend charming relief to the otherwise plain sides of the cabinet. The horn is colorful and has interesting contours. Machines such as this were frequently unmarked, and sold under a host of rather pretentious names. *Courtesy of Sam Saccente.* (Value code: G)

3-81. When it first began selling Pathéphones, the Girard firm traded as J.Girard and Company. In 1909, the firm was re-christened Girard and Boitte. This *trompe l'oeil* decal, meant to imitate the metal ID plates (right down to the *faux* screw heads) that were affixed to the cabinets of most Pathéphones, bears the later designation of the mail-order firm. *Courtesy of Michael and Suzanne Raisman.*

3-80. Early in the twentieth century, the Girard company of France began selling Pathé cylinder talking machines by mail order. After Pathé started making disc machines (introduced 1906), Girard applied the same marketing techniques to disc Pathéphones. The machine illustrated here was essentially a Pathéphone Model "No.4" housed in a cabinet specially designed for Girard. The customary discus-thrower decal Pathé placed on its disc machines from about 1908, and the Pathé-attributed soundbox leave no doubt as to the manufacturer. *Courtesy of Steve and Theresa Sposato.* (Value code: G)

3-82. The machine shown here, known as the Cameraphone, resembles a Type "BC" Graphophone equipped with a mandrel for 5" diameter cylinder records, but it did not supply entertainment of the domestic kind. The Cameraphone was the main component in an early attempt to synchronize sound with moving pictures. The National Cameraphone Company was incorporated on March 26, 1907, by James A. Whitman and Eugene E. Norton, who was formerly a mechanical engineer at American Graphophone. The Cameraphone factory, managed by Norton, was located in the Graphophone's home turf, at 423 Waters Street, Bridgeport, Connecticut. The superintendent, A.A. Stevenson, had been foreman of the tool room at the Graphophone plant. The link to American Graphophone couldn't have been stronger, and it is possible that this machine was modified at the Graphophone factory to the Cameraphone Company's specifications. The first public demonstration of the Cameraphone occurred on June 10, 1907 at Oscar Hammerstein's Paradise Roof Garden in New York City. Norton was awarded a patent on a clutch mechanism to adjust synchronization in March 1908. More successful exhibitions soon followed and, by April 1908, Edison's industrial spies were attempting to infiltrate the company. It was discovered that the Cameraphone system used a Pathé camera, which infringed a patent acquired by Edison's Motion Picture Patent Company. Legal steps were taken by both sides, but the Cameraphone apparently sputtered out within two years. The eventual failure of the Cameraphone system probably had little to do with sound quality. The system's Higham reproducer was an efficient loud-speaking apparatus, and, coupled with the higher surface speed of a 5" cylinder, the playback volume should have been impressive. Like other cinematic efforts that preceded it, the Cameraphone probably suffered from unsatisfactory synchronization, a problem which was not overcome for another 15 years. *Courtesy of Douglas DeFeis.* (Value code: VR)

3-84. The vibrations of the steel needle were greatly augmented by compressed air passing through the "Auxetophone" soundbox. A large horn insured proper distribution of the sound. *Courtesy of the Sanfilippo collection.*

3-85. Within the cabinet of the "Auxetophone" was an air compressor run by an electric motor. The example shown is one of a very few of these machines retaining its original compressor unit. *Courtesy of the Sanfilippo collection.*

3-83. In 1904, Gramophone and Typewriter, Limited of Great Britain introduced a loud-speaking disc machine known as the "Auxetophone," which was most commonly available in a version with a large external horn, although an enclosed cabinet model was later catalogued. The device's compressed air method of sound amplification was the invention of C.A. Parsons. The external-horn "Auxetophone" was licensed to the Victor Talking Machine Company, and released under the Victor brand name in May 1906. The cabinet of this example was the more ornate of the two external-horn styles Victor offered. The price was $500.00. *Courtesy of the Sanfilippo collection.* (Value code: VR)

3-86. Klingsor was a brand of disc talking machine which enjoyed a remarkably long life. Developed in Germany during the first decade of the twentieth century (patented by Heinrich Klenk, No.899,491, September 22, 1908) it eventually became a product of British manufacture as late as the mid-1920s. A great variety of styles can be found, but the basic premise was constant: tuned strings were drawn across the mouth of an internal horn, with supposedly harmonic results. The vertical orientation was usually maintained in Klingsor machines, although the shape of the cast metal harp varied, and many models incorporated colorfully glazed doors. The particular example shown here, which has had some minor decorative modifications, is from the earliest period of Klingsor manufacture: illustrated in the 1907 catalogue as Model "No. 125," selling for DM 125. The playing compartment was accessed by pulling the brass handle above the name plate. A conventional "Victrola" style table model also appeared under the Klingsor name, with low profile, two sound-modifying doors at the front, and a merely rudimentary harp. *Courtesy of Sam Saccente.* (Value code: F)

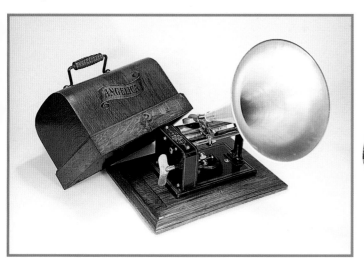

3-87. Part of the "German wave" of the 1907 period was this "Angelica" cylinder talking machine made by Excelsior and marketed in Great Britain. Many manufacturers at this time subscribed to the theory that the more sublime the brand name, the more successful the phonograph. *Courtesy of Steve and Ellie Saccente.* (Value code: G)

3-88. In November 1906, an advertisement appeared in the *Talking Machine World* offering shares of stock in the Multiphone Operating Company of New York City. Boasting 4% monthly return on investments of stock at $100 per share, the ad estimated an approximate annual profit for each operating Multiphone of $363.50. "Practically every penny earned is for dividends, for there are no salaries; there are no expenses; there are no leaks. You couldn't think out a cleaner, squarer, more attractive business… If you care to get in on the ground floor of the richest, juiciest business chance that ever came your way—look into this one. We don't want a cent from a blindfolded man—the further back you are from Missouri the better we will like it." Yet, by May 1908 a receiver had been appointed for the bankrupt Multiphone Operating Company. The following month, panicked stockholders organized a protective organization in an effort to save the company's assets. The counsel for the receiver of the Multiphone Company told stockholders that he believed "…[I]f the operating companies were put in the hands of an efficient management and dividends delayed until the earnings justify them, the companies would be able to get on their feet again. It is said that more than $1,000,000 had been put into stock…by thousands of investors." An advertisement appeared in the December 1908 issue of the *Talking Machine World* announcing in bold print: "MULTIPHONES TO BE SOLD." "The Multiphone has a large magazine wheel carrying 24 records. Either Edison, Columbia or indestructible [sic] records may be used. The instrument is purely automatic, and operates for a nickel. A spring motor supplies the power. One winding is sufficient to reproduce from 20 to 25 records. The Multiphone can also be adjusted in a moment to automatically play all of the 24 records, passing automatically from one to the next without any attention other than winding at the start." The financially-strapped Multiphone company was reorganized and continued, and over the life of the brand several models were produced of which this one was by far the grandest. The machine stands 7' 2" tall, 3' 6" wide and 18" deep. *Courtesy of the Sanfilippo collection.* (Value code: VR)

3-89. Pathé maintained a huge record catalogue with discs in a considerable number of diameters. As the discs became larger, so did the machines to play them. This stunning "Concert," circa 1909, with aluminum horn in *torsadé* (twisted) style, could accommodate records of a formidable 50cm (approximately 20") in diameter. There was a series of Pathé "Concert" machines of which this instrument was designated Model "A." Models "C" through "E" were increasingly large wooden closets which completely enveloped the machine shown here. (Value code: F)

3-90. The growing popularity of the talking machine during the first decade of the twentieth century dealt a serious blow to manufacturers of music boxes. Even the venerable Regina Music Box Company of Rahway, New Jersey, which produced music boxes that played interchangeable metal discs, was not immune. In 1904, Regina introduced the first of its "Reginaphones." These were music boxes with disc talking machine components attached in order to give the purchaser both types of entertainment. The talking machine parts were manufactured by the American Graphophone Company, and usually added $25.00 to the cost of the box. Shown is Reginaphone Style "172," adapted from the Style "72" 12 1/4" disc, single comb music box. This outfit cost $75.00 (in 1908) and was produced between 1907 and 1916. *Courtesy of Charles McCarn.* (Value code: C)

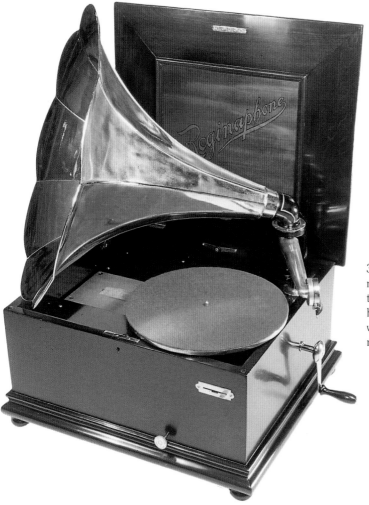

3-91. A Reginaphone, Style "113," circa 1908. This instrument played 15 1/2" metal music box discs when the horn, tone arm assembly and turntable were removed. Style "113" had been in production since 1904, when it was supplied with the earlier "front-mounted" variety of phonograph equipment. *Courtesy of the Sanfilippo collection.* (Value code: C)

3-92. Beginning in 1905, the earliest coin-slot Regina cylinder talking machine was a penny-operated table-model, playing six two-minute cylinders (later available in a four-minute version). The small flower horn on this example, resembling the later Repp Vitaphone horn, was a greater embellishment than the straight brass or nickel-plated horns usually associated with coin-ops. *Courtesy of the collection of Howard Hazelcorn.* (Value code: VR)

3-94. When opened, the machine reveals its secrets: a bracket, elbow, arm and reproducer directly adapted from an external-horn Graphophone. At this early stage in the development of the internal horn machine, the demarcation between old and new technology was far from crisp. *Courtesy of the Sanfilippo collection.*

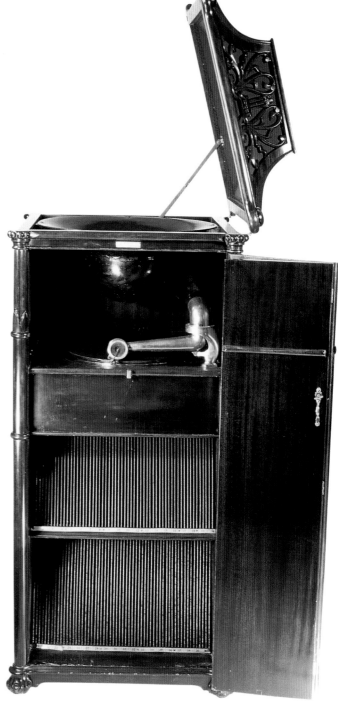

3-93. The Columbia "Symphony Upright" which, in 1908, followed upon the heels of the "Symphony Grand," was a pricey ($200.00) machine in a monumental style that was shared by Hawthorne and Sheble's "Starola" and Keen-O-Phone's "XXX." The tall body of the "Symphony Upright" concealed a metal-bell horn directed upwards to allow the sound to exit through the decorative grille-work. The little celluloid plate visible at the top of the cabinet, which was also used on the "Symphony Grand," listed the only reference to Columbia on the cabinet of the machine. *Courtesy of the Sanfilippo collection.* (Value code: E)

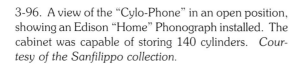

3-96. A view of the "Cylo-Phone" in an open position, showing an Edison "Home" Phonograph installed. The cabinet was capable of storing 140 cylinders. *Courtesy of the Sanfilippo collection.*

3-95. The Herzog Art Furniture Company of Saginaw, Michigan introduced this cabinet "No. 740" in April 1908. Its purpose was to allow the owner of an external-horn cylinder machine to "convert" it into a more Victrola-like, concealed horn instrument. Available in either Mahogany or Golden Oak (shown), this cabinet, advertised as the "Cylo-Phone" (pronounced *sillo,* as in *cyl*inder) seems to have disappeared by early 1909. *Courtesy of the Sanfilippo collection.* (Value code: E, with machine)

3-97. It's anybody's guess what wonderful instruments might have emerged from the Hawthorne and Sheble Manufacturing Company, had it not been brought to an untimely end by Victor's legal department. In December 1908, at the peak of its innovative powers, H&S introduced the "Starola." Drawing the name from the existing line of H&S "Star" external-horn disc machines introduced in mid-1907, the "Starola" was an internal-horn model cast in the "monumental" style of the earliest Grafonolas and the Keen-O-Phone "XXX." Shown is Starola Grand, Model "250," which sold for a lofty $250.00. It was available for less than a year before H&S collapsed under Victor's permanent injunction. *Courtesy of the Sanfilippo collection.* (Value code: F)

3-98. The interior of the Starola Grand, Model "250" displayed the same soundbox and tone arm used on H&S external-horn machines. Patented features of this arm were the "Yielding Pressure Feed" (a spring-activated pseudo-feed device) and a built-in volume control. Imitating the orientation of external-horn models, the Starola internal horn was located above the playing mechanism. Worn steel needles fell down the chute under the soundbox into a receptacle accessible from the record storage compartment below. To the left is an H&S multi-compartment needle tray. *Courtesy of the Sanfilippo collection.*

3-99. The Victrola "XX" ("the Twentieth") was introduced March 1, 1908 as the premier offering of the Victor Talking Machine Company. Originally selling for $300.00, the "XX" featured distinctively figured Laguna mahogany applied at 45-degree angles, and gilded ornamentation. The fact that this particular Victrola "XX" had no gilding suggests that such decorative touches were not universally admired. The buying public evidently recognized that the "XX" was essentially a slightly dressed-up "VTLA"/Victrola "XVI," and by December 1908, the price of the remaining 275 machines in stock was reduced to $250.00. Only 12 months after its introduction, the Victrola "XX" was officially discontinued. Not until 1915 would Victor catalogue another $300.00 Victrola (the "XVIII"). *Courtesy of Walter and Carol Myers.* (Value code: E)

3-100. John Gabel (1872-1954) was a fourteen year-old Hungarian when he immigrated to America in 1886. He spoke no English, and childhood illness had limited his formal schooling to only two years. He worked odd jobs while staying with a brother in Philadelphia. At age sixteen, Gabel moved to Chicago, where he found a job assembling adding machines. A born machinist, he worked for several firms which manufactured coin-operated gambling machines. By 1898, Gabel had formed his own concern: the Automatic Machine and Tool Company, which boasted three employees. Gabel's first design, a mechanical floorstanding gambling machine, was an immediate success, and by 1900 Gabel employed 50 men. For Christmas 1903, Gabel brought home a Victor, and became interested in designing an automatic talking machine. By the spring of 1905, his prototype was ready. Gabel filed his first patent on the "Automatic Entertainer" on February 26, 1906, and began marketing the device. Victor rapidly brought suit for patent infringement, and despite repeated adverse decisions, kept up the pressure until 1912, when the "Berliner Patent" expired. "Gabel's Automatic Entertainer" was the world's first selective disc mechanism, offering a choice of 24 records. The soundbox was driven across the record by a feedscrew, and the instrument employed a slug detector and coin counter, innovative features for the time. One turn of the crank wound the motor and changed both the record and the needle. Several variations exist, since this instrument was in production until 1929, an incredible 23 years. The example shown dates from 1909-1910. The crank was inserted below the right-hand knob at the front of the cabinet. *Courtesy of the Sanfilippo collection.* (Value code: VR)

3-101. The Regina Company of Rahway, New Jersey, famous music box manufacturers, made numerous sorties into the talking machine business during the first two decades of the twentieth century. One of the most successful was the introduction of the Regina "Hexaphone" coin-operated cylinder talking machine. The basic mechanism appeared in several forms over a period of about 15 years. Shown is a two-minute version (Style "101" circa 1910). It should be recalled that even though four-minute cylinders were introduced at the end of 1908, a large trade in the two-minute variety continued for several years. A six-record carousel was loaded with cylinders, any of which could be selected by the customer for a nickel. *Courtesy of the Sanfilippo collection.* (Value code: B)

3-102. The Pathé "Omnibus" was a very unpretentious model sold inexpensively, circa 1909. It incorporated all the customary elements of Pathé back-mounted disc machine design, but on a modest scale. *Courtesy of Michael and Suzanne Raisman.* (Value code: G)

3-103. Hawthorne and Sheble Manufacturing Company of Philadelphia produced a good many "client" or contract disc talking machines, as well as the company's own Star line. H&S instruments commonly appeared under the "Busy Bee," "Yankee Prince" and "Aretino" brands. Here we see a much less well-known mark: "King." The customary components which distinguished H&S machines, including tone arm with the internal features of volume control and "Yielding Pressure Feed," and the convenient, side-mounted used-needle receptacle, were present in this machine. A design characteristic of Hawthorne and Sheble was the grain-painted flower horn with raised panels, which the company also made in an elongated version, to fit cylinder instruments. (Value code: F)

3-104. Contained in this 1909 Columbia "Climax" are elements which evoke the stories of America's cleverest independent entrepreneurs. In general appearance, the machine resembled a model that had been offered by the Hawthorne and Sheble Manufacturing Company of Philadelphia in 1908. There was a legitimate reason for this: the "Climax" was partially constructed of former H&S inventory. In late 1909 when the "Climax" appeared, Hawthorne and Sheble had recently ground to a smoldering halt under the weight of patent infringement litigation instigated by Victor. Though court proceedings would drag on through the following year, H&S was effectively out of business. A letter sent by Hawthorne and Sheble to Mr. W.R. Howie of Beebe Plain, Vermont, postmarked May 28, 1909, had optimistically included a dealer's contract. More tellingly, Form No.214 was also enclosed, which read in part, "We are closing out all our stock of New Flower Horns for Talking Machines, as we are going to discontinue manufacturing horns except where employed on our own machines. Also our surplus stock of Carrying Cases and Needles…" H&S was putting a good face on a very desperate situation, and trying to raise capital. We may also surmise that the Hawthorne and Sheble Manufacturing Company, like Atlantic Talkophone before it, attempted to survive by selling parts to other firms. The cabinet, back-bracket and motor control of this "Climax" were H&S parts. A slip-fit elbow incorporating traditional Columbia threading secured a Columbia "No. 1" mahogany horn. The motor was the same triple-spring Columbia model that appeared in such machines as the (late) "BI" and "BII." The tone arm was a custom-designed, ball-and-socket affair adapted to the H&S back-bracket. A knurled cap may be seen at the top of the ball-joint, which is where the former H&S tone arm (patented by Mr. Sheble) once attached. This machine is dated on the underside, October 14, 1909. It's worth noting that one of the proprietors, Horace Sheble, later found work with the American Graphophone Company (Columbia). Furthermore, a mahogany cabinet nearly identical to the one shown here appeared a short time later fitted out as a Keen-O-Phone. Keen-O-Phone of Philadelphia, whose personnel were related to H&S, might also have also obtained surplus H&S cabinets. Despite its triple-spring motor, the "Climax" was a patchwork machine, bearing no attribution to Columbia. It was intended to look expensive, but sell inexpensively. Columbia also pressed a series of inexpensive "Climax" discs. Of course, the name was a revival of the "Climax" appellation with which Columbia had entered the disc record business in 1901. (Value code: E)

3-105. The growing success of the Victor-Victrola prompted the company to introduce its first table model Victrola in July 1909. Sometimes called the "Baby Victrola" in contemporary advertising, the Victrola "XII" retailed for $125.00. Perhaps the most noticeable feature of this model was the extremely small horn opening, accessed through correspondingly squat "sound modifying" doors. It was immediately apparent that, despite most of the same mechanical components found in the Victor "VI" and the "VTLA," the Victrola "XII" was incapable of comparable performance due to its radically reduced horn size. Victor advertising valiantly proclaimed the merits of the new horizontal "sounding boards" within the mouth of the horn, but few buyers took the bait. At a price $25.00 more than the top-of-the-line external-horn Victor "VI," the Victrola "XII" was no bargain. Victor designers may have understood that the cabinet was rapidly becoming the principal talking machine selling point, but the Victrola "XII" demonstrated that there were limits to what the public would tolerate. In an effort to boost sales, additional frieze work was added to the cabinet in January 1910, but the model remained unpopular. By September, Victor threw in the towel and discontinued the Victrola "XII." Hereafter, Victrola horn sizes would be roughly three times that which was offered in the "Baby Victrola." The "sounding boards" would become a standard Victrola characteristic, perhaps the only lasting contribution of this short-lived model. (Value code: G)

3-106. A seasonal Victrola advertisement from 1909. Note the Victrola "XII" in Santa's sleigh. *Courtesy of Alan H. Mueller.*

3-107. The Victor factory in 1910.

3-108. A Victor Talking Machine Company stock certificate (unused). *Courtesy of the collection of Howard Hazelcorn.*

3-109. In 1909, William H. Hoschke patented a disc talking machine (No.948,327) with a mechanically tracking soundbox, and began marketing it as the "Sonora." Sonora instruments were brashly promoted at a time when the rest of the disc talking machine market was cowering in fear of Victor and Columbia (and the powerful patents they wielded with impunity). Victor sued claiming the tracking soundbox was a ruse. In late 1910, a federal judge agreed, asserting that the Sonora feed device was not absolutely crucial to the operation of the machine. Sonora stood its ground. It attacked the Victor patent under which it had been charged, and some months later the same judge ruled that "Berliner Patent" No.534,543, the jewel in Victor's legal crown, was no longer valid because it should have already expired (due to the prior expiration of a related patent). Although Victor successfully appealed the decision, Sonora celebrated by introducing a line of machines *sans* feed devices. Sonora, through a relationship with the famous Swiss music box firm of Palliard, would create some of the highest-quality and most interesting talking machines in the disc market, and later be reborn as a company whose signature *bombé* styling became renowned. This very early Sonora model, with characteristic "breadbox" cabinet, shows the mechanical feed conduit with tracking soundbox along the rear of the playing compartment. *Courtesy of the Domenic DiBernardo collection.* (Value code: F)

Opposite page: 3-110. An advertisement for the mechanical-feed Sonora refers to it as the "Sonoraphone" and emphasizes vertical cut (sapphire) records, which would be the refuge of the smaller record companies during the 'teens. In 1910, only Keen-O-Phone of Philadelphia was on the verge of producing vertical-cut discs in the United States, so the records referred to in the ad copy must have been foreign imports. *Courtesy of the Domenic DiBernardo collection.*

3-112. Shown is the upper mechanism of a "transitional" Edison Amberola "IA/IB." The "IA," a belt-driven, two-and-four minute machine, was discontinued late in 1911 when replaced by the "IB," a direct-gear drive, exclusively four-minute instrument. An unusual combination was the maroon-painted mechanism housed in an oak cabinet. Oak cabinets were usually supplied with works finished in "gun metal." Certain inconsistencies in this mechanism are remarkable. Although this is clearly a "IA" upper works (and the cabinet bears a "IA" ID plate), the machine is *four minute only.* The gold knob for changing between two-minute and four-minute cylinders is absent from its accustomed location at the rear of the casting. Furthermore, there is no hole where it might have been located. Neither is there evidence of the speed control and indicator usually found in the left rear corner of the bedplate. Instead, the unusually large speed control has been relocated to the left front corner of the casting. These unconventional alterations suggest that there may have been a few unfinished "IA" upper castings remaining when the "IB" was inaugurated. Although the larger manufacturers usually eschewed "patchwork" machines, Edison is known to have hurriedly assembled instruments in the early Amberola line to expedite shipments. *Courtesy of the Sanfilippo collection.* (Value code: E, ordinary Amberola)

3-111. Upon its introduction in late 1909, the Edison Amberola "IA" was available in Oak or Mahogany. By July 1910, the machine could be ordered in Circassian Walnut, as shown here. This option boosted the price of the Amberola to $250.00 The external metal parts of this particular machine are gold-plated, while the horn neck and reproducer (Model "M") are oxidized bronze, and the upper works are gun metal finish. Note the cloven feet on this example, one of several variations seen in these cabinets. (Value code: VR)

3-113. This "Maestrophone No. 205" talking machine (with "Maestoso No.2" soundbox), produced around 1910 by the prodigious Swiss firm of Paillard, used a hot air motor to power the turntable. An alcohol lamp heated the air, which ran a turbine. The cabinet included two glass panels to satisfy the curiosity of those being entertained. What appears to be the bell of a horn emerging from the left side of the cabinet is actually an exhaust vent. Because of the ever-present danger of fire, the hot-air motored talking machine was more a novelty than a practical alternative to spring or electric power. *Courtesy of Ray Phillips.* (Value code: VR)

3-114. The Pathé company of France had a way of making eccentric jewels of talking machines. This little instrument, with its turntable only 5 7/8" in diameter, delights us with its charm. It was most often sold under the name "Jeunesse" (Youth). Pathé has emphasized the horn rather than hidden it, although the "Jeunesse" was sold circa 1910 when visible horns were about to become *passé*. *Courtesy of Steve and Ellie Saccente.* (Value code: G)

3-115. The Hiller clock was one of the rare breed of time-keeper which was intended to speak the hours, and thereby eliminate the necessity of reading the face. Various reasons were suggested for why such a device would be more than a mere novelty. However, the complexity of the mechanism and the need for it to remain precise doomed the true talking clock to the realm of gimmick. *Courtesy of Jean-Paul Agnard.* (Value code: VR)

3-116. The workings of the Hiller clock can be seen in this open view. Note the sprocketed pulleys that moved a loop of celluloid film (without photographic emulsion) beneath a rather conventional disc talking machine reproducer. On the film was a vertical recording of the hours and quarter hours. The sound was emitted through a small horn aimed at a grille in the top of the wooden case. The Hiller clock was manufactured in Germany, circa 1911, under the patents of B. Hiller. The cabinet measures 16 1/2" high. The width of the film between sprocket hole centers is 40mm. *Courtesy of Jean-Paul Agnard.*

3-117. An Edison "Idelia" model "D-2" ($125.00) fitted after August 1910 with a Spruce "Music Master" cygnet horn ($15.00) and a "Diamond B" reproducer. The "Idelia" was the premier Edison external-horn cylinder Phonograph until the introduction of the "Opera" in November 1911. *Courtesy of the Sanfilippo collection.* (Value code: B)

3-118. Out of the wood-cased "Edison Business Phonographs" of the twentieth century's first decade grew the all-metal "Ediphone." Under this designation, the Edison commercial phonograph would survive for decades. By the mid-'teens, "Ediphones" looked like this: compact, electrically driven, table model dictation units mounted on thin legs. The ordinary color was uninspiring black. This highly unusual example, finished in the same rich maroon color that distinguished the Model "D" "Gem" and the interior works of the Amberola "1A," suggests how much more appealing the drab "Ediphone" might have been made to the eyes of a collector. *Courtesy of Walter and Carol Myers.* (Value code: I)

3-119. A 1912 "guide" offered by Edison, intended to aid in the selection of a dictating machine. The success enjoyed by Columbia's "Dictaphone" clearly prompted this comparison of features.

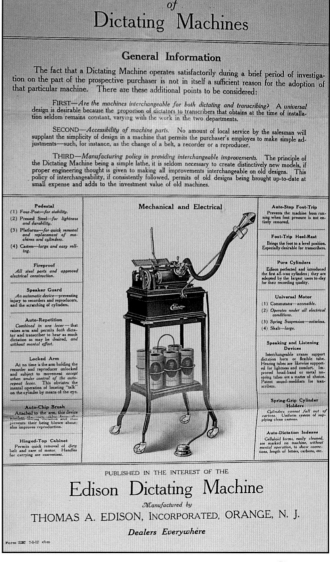

"Pardon the Interruption—
— I must sign my mail before closing time"

3-120. An advertising flyer for Dictaphone begins by illustrating an awkward business moment. The Dictaphone was Columbia's cylinder office machine, which had begun during the first decade of the twentieth century as the "Commercial Graphophone."

3-121. The solution is summed up in the ad copy: "Use the Dictaphone—that's the answer!"

That Nightly Rush to "Get the Mail Out"

"ARE your letters ready, Mr. Jones?" Time after time your stenographer asks that question just before "quittin' time," with that anxious, half-scared look on her face.

What can you do? You hate to be interrupted. Perhaps you're talking with a customer, or discussing some important matter with a department head. But the letters must go out, and you can hardly ask the girls to stay overtime.

What can you do? Use the Dictaphone—that's the answer! Then dic-

Reach for your telephone, call up The Dictaphone, and arrange for a demonstration in your own office, on your own work.

3-122. For commercial purposes, Edison offered this hand-driven "Universal" record shaver circa 1908. Note the mandrel to accommodate the 6" long business cylinders. Since all Edison home entertainment machines had lost their built-in shaving attachments by 1908, the "Universal" was also suggested for the home, though it is unlikely that it saw much private use. *Courtesy of the Domenic DiBernardo collection.* (Value code: I)

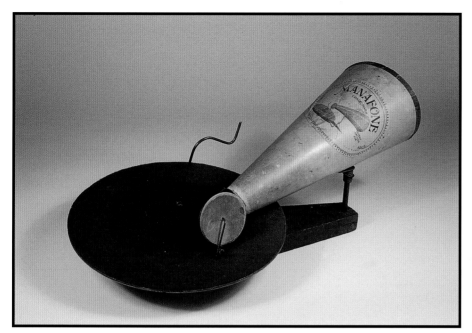

3-123. This "Manafone" was a simple, hand-driven talking machine with no markings other than the attractive horn label. The soundbox was an anachronism, resembling that of a Berliner Gramophone from an earlier period. The turntable is 9" diameter. (Value code: H)

3-124. This unmarked machine, while clearly designed to sell at a low price, was nevertheless substantially built. Equipped with a 7" turntable, this machine appears to have been manufactured in the first decade of the twentieth century. (Value code: H)

3-125. The "Wondophone" talking machine, like the "Manafone," was probably given away for selling magazine subscriptions or soap. The flimsy, pressed-metal base failed to hold the 8" turntable properly, and the original owner of this device may well have been disappointed by the fruits of his efforts. Phonographs of this ilk demonstrate the enormous popularity of talking machines in the early twentieth century, and the resultant market for even the cheapest of the genre. (Value code: H)

3-126. This little internal-horn machine, sold under the "Lakeside" brand by Montgomery Ward of Chicago, was typical of the client machines produced by Columbia, circa 1911. Practically identical instruments appeared under a variety of names in Chicago alone, including "Standard," "Harmony," "Symphony" and "Aretino." Shown is the Lakeside "Gold Medal," which sold for $9.90. *Courtesy of Robin and Joan Rolfs.* (Value code: I)

3-128. A "Cortina Language Record" box, depicting on the label a distinguished-looking gentleman studying with a Cortinaphone. "Made from our own moulds, U.S. Everlasting Records, which are indestructible."

3-127. By the time the second decade of the twentieth century began, the talking machine business in America had compressed to an impenetrable kernel containing three firms: Victor, Columbia and Edison. Nearly all the independents who had begun life during the previous ten years had folded, and the U.S. Phonograph Company of Cleveland, Ohio, was one of the last to go. Beginning in May 1910, the U.S. Company made cylinder talking machines which were cleverly-designed and well-built. It also adapted its external-horn "Rex" model for use in home language teaching. Like all U.S. Phonograph Company machines, the Cortinaphone, sold by R.D. Cortina Company, New York, included both two and four minute gearing. Of the variations of the external-horn "Rex" (U.S. "Rex," Lakeside "Rex," and Cortinaphone) only the Cortinaphone employed a wooden cabinet cover. By the end of 1913, the U.S. Company, exhausted from long legal harassment by Edison, ceased operation. (Value code: G)

CHAPTER FOUR

1912-1919

The 'teens were the salad days of the internal-horn disc talking machine in the United States. It was a time when dozens and dozens of small firms offered instruments that were sometimes beautiful, sometimes peculiar; sometimes brilliantly conceived, sometimes self-consciously contrived; sometimes uninspiringly ordinary, but always representative of the energy which abounded during America's second decade of the twentieth century.

On February 19, 1912, Patent No.534,543, the "Berliner Patent," officially expired. For over ten years it had been the terrible swift sword with which the Victor Talking Machine Company had decimated the independents. The following day, the flowering of small disc talking machine enterprise in the United States began anew. The Berliner Patent had been Victor's foremost weapon against the competition because it addressed a practically inescapable feature of disc talking machine design: the use of the record groove to carry the soundbox across the disc during play. In order to circumvent this fundamental premise, independents had attempted all manner of contrivance and subterfuge. Complicated mechanical tracking systems were expensive; simple spring-loaded feed devices were ineffective. Only the Keen-O-Phone Company of Philadelphia managed to stagger into the post-Berliner Patent world, from the minefield that had been the recent American disc talking machine industry. Of the independents to commence production in 1912, the Vitaphone Company of Plainfield, New Jersey, was one of the first.

A BRIEF HISTORY OF THE THEORY BEHIND C.B. REPP'S VITAPHONE

The idea of transferring sound vibrations from a reproducing point to a resonating device through a solid medium was employed in too many offbeat talking machines to describe the concept as a fluke. Conventional acoustic talking machines conveyed the sound energy from the reproducer (or soundbox) to the amplifying horn through a column of air contained in a hollow tube or arm. However, certain small companies chose to avoid this "hollow conduit" method of sound transference, which had a variety of patents associated with it, in favor of a walk on the weird side.

The first example of unconventional sound transference can be found in Edward H. Amet's Echophone of 1895. This diminutive device was the very first low-priced cylinder talking machine (at $5.00). It was briefly marketed and accomplished little more for its inventor than to get him sued for patent infringement by the American Graphophone Company. The mechanism was simple, employing a clock motor to run the record mandrel. The reproducing stylus was formed out of the end of a long glass tube. The tube was attached to a resonating chamber made of wood and rubber. From this chamber sound would emerge into a set of rubber listening tubes, or a small, conical horn. Amet described the passage of the sound energy from the stylus, along the tube and into the resonator as "molecular vibration." The glass tube was hollow, but since it was mounted on a wooden peg at the top of the resonator, there was no movement of air between tube and resonating chamber. It was the vibration of the glass itself, not the air which might be contained within, which carried the sound impulses. This arrangement put the short-lived Echophone squarely in the league of "solid medium" sound transference.

In 1897, the United States Talking Machine of J.N. Brown appeared. Another low-priced machine ($3.00, later cleared out at 98 cents!), it was designed to play 7" diameter Berliner disc records. The record sat on a small turntable, rotated by hand. A strip of wood held a steel needle at one end and was attached at the other end to a set of rubber listening tubes. As the needle tracked across the record, the simple wooden arm carried the sound vibrations to the tubes. The air contained in the

ear tubes acted as the resonating space which allowed the sound to be heard. Mr. Brown, apparently less gregarious than Mr. Amet, stated in his patent application (granted as patent No.653,654) that he was "not prepared to explain the phenomena" by which his apparatus functioned. Once again, this "solid medium" transference device flickered only briefly in the marketplace.

Over the decades, the elements of sound transference initiated by the Echophone and the United States Talking Machine periodically reappeared in acoustic phonographs. Even long after the advent of electrical reproduction, a cheap, hand-driven, children's record player could be found, employing a modification of the same solid wood sound transference which had originated in the 1890s. However, in one instance the genre was elevated to the level of a genuine musical instrument: the Vitaphone of the early 'teens.

4-1. An 1896 advertisement showing Echophones being closed out as premiums for *Leslie's Weekly*, and an 1897 advertisement showing the United States Talking Machine.

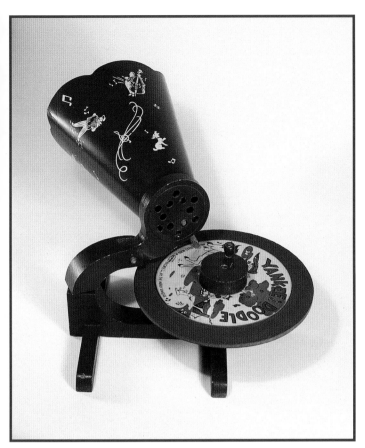

4-2. This simple children's acoustic phonograph from the 1950s employed a variation of the "solid medium sound transference" first developed in 1890s talking machines. A steel needle is attached to the wooden dowel protruding from the base of the horn. The sound vibrations pass through this dowel to a diaphragm above, rather than through a hollow arm as was customary in most other acoustic phonograph designs. *Courtesy of Walter and Carol Myers.* (Value code: I)

THE VISCERAL VITAPHONE

Clinton B. Repp was the force behind the Vitaphone. In 1899, he had been involved with marketing a failed disc machine known as the Vitaphone, from which he borrowed only the non-trademarked name. In 1909, he had applied for a patent which was issued on September 19, 1911 (No.1,003,655) for a wooden sound-conducting arm, stationary reproducer and amplifying horn. Repp would go on to receive at least four German patents for his acoustic talking machine designs. According to an overly-optimistic and rather colloquial *Talking Machine World* in April 1913, "…and as patents secured in Germany mean a whole lot, it shows the broadness of the different principles used in the construction of the Vitaphone." Despite the reporter's faith in foreign patents, it was Repp's American patent upon which the Vitaphone Company was founded.

As in numerous other stories of independent talking machine enterprise, things did not begin propitiously. On November 18, 1911, the Victor Talking Machine Company was granted an injunction against the fledgling Vitaphone Company that prohibited "manufacturing, selling or using a machine called the Vitaphone and manufactured under the Repp patent, on the grounds of infringing the Berliner patent." It is comforting to know that Victor's legal department was still hard at work flagellating the competition with the Berliner Patent (No.534,543) within a few months of the patent's already disputed expiration date. Because of the "eleventh hour" circumstances, the injunction included this qualifying statement: "Provided, however, that the defendant [Vitaphone] may advertise, if it chooses to do so, that after February 19, 1912 [the official expiration date of the Berliner Patent]… it will manufacture and supply [talking machines to] the trade."

During 1912, the Vitaphone Company occupied an experimental plant in Plainfield, New Jersey. By August, an adjoining four-story factory was planned. The firm announced that it would manufacture a line of models priced from $15.00 to $185.00. In October 1912, the *Talking Machine World* reported that all Vitaphone machines would be "hornless" (internal-horn), with the accommodation for quickly appending an external horn if greater volume were needed. Even though it was widely recognized that internal-horn machines produced inferior sound quality, one may infer from this statement that public taste was already well established in regarding the external horn as old-fashioned. Vitaphone's secretary and general manager, longtime talking machine salesman H. N. McMenimen, concurred: "While the public seems to demand a hornless constructed machine, yet most of us in the trade realize that the horn presents a greater detail and volume of the reproduction."

Production of Vitaphones was underway by the end of 1912. The introductory line consisted of one low-cost external-horn model, followed by four internal-horn machines of ascending price. Since the idea of a "convertible" internal/external-horn model was promoted in Vitaphone catalogues, some effort must have been made to sell buyers, especially of the company's smaller internal-horn table models, an optional external horn and connecting elbow. In addition to a rather squat, red-painted metal flower horn, Vitaphone offered a wooden horn. During the 1912-1915 period, the wooden horn, previously a sign of quality and high price, devolved to the purpose of "dressing up" inexpensive talking machines. Although Vitaphone planned to introduce wooden horns which were "solid, veneered, wood-pulp and solid spruce," only a solid oak horn was actually sold (the "Baby Music Master" manufactured by Sheip and Vandergrift of Philadelphia).

Vitaphone's early patent conflict with Victor (relating to the tracking of the needle across the record) seems ironic when one considers the spring motors with which Vitaphones were equipped. These motors were virtual copies of the reliable, worm-driven devices which had powered Victor machines since 1906 (primarily models "II" and "III"). There was nothing to suggest that the Victor company objected. However, the heart of the Vitaphone was not in the motor, nor in the cabinetry, but the "solid medium" transference of sound.

A 1913 Vitaphone catalogue described Repp's reproducing system (with a liberal interpretation of the laws of physics):

An elementary knowledge of tone appreciates the fact that the most natural and musical tones are produced through the medium of WOOD rather than metal. The violin, the flute, the organ pipe, the 'cello are but common types which prove the soft, mellow tones of wood which improve as time goes on. The Vitaphone is constructed along this distinctive line, and owes much of its distinctive tone to the use of the patented WOODEN ARM which carries all the sound waves from the delicate needle to the patented STATIONARY SOUND BOX which is not swaying with the wave of the record....One can feel the every tone vibration throbbing through this wooden arm as it is carried to the stationary sound box where it is reproduced, and then instead of being diverted downwards, it is allowed to float UPWARDS as is natural with sound waves.

Included in the final reference were Vitaphone's internal-horn floor models, which housed the amplifying chamber in the lid of the cabinet, such as the Keen-O-Phone Company was doing contemporaneously.

Mr. McMenimen, waxing classical, echoed Edward Amet's phraseology when he told the *Talking Machine World* in October 1912: "...the Vitaphone is taking advantage of the knowledge [of]...vibrating tones by a *molecular displacement* [emphasis ours] of the wood and being applied to talking machine reproduction. If one were to take a string of a violin with a tin body...nothing would be secured but a very nasal metallic screeching sound, whereas the constant displacement of molecules of the wood in the body [of a violin] gives that sonorous, sweet tone that has even caused destinies to fall."

The solid length of "violin wood" through which the Vitaphone's sound vibrations passed was weighted to make the needle exert considerable pressure against the record, and thereby release as much physical energy from the groove as possible. The mica diaphragm of the stationary soundbox was kept under tension, which aided the transference of the vibrations from the arm along a short length of gut. In fact, both laterally and vertically-generated vibrations would pass equally well through the wooden arm. This meant that disc records of either type could be played without the necessity of clunky pipe-fittings, common among conventional disc talking machines.

In 1913, the Vitaphone line consisted of an external-horn model ("No.15," $15.00), a small internal-horn table model ("No.18," $18.50; mahogany, light or dark oak or mission finish), a slightly larger internal-horn table model ("No.25," $25.00; quartered oak, light or dark finish), a horn-in-lid model on high legs ("No.50," $50.00; mahogany, dull rubbed finish; fine quartered oak, light or dark finish), a horn-in-lid enclosed cabinet model in Sheraton style ("No.75," $75.00, mahogany, dull finish; light, dark mission finish, quartered oak), a larger horn-in-lid enclosed model in similar style ("No.100," $100.00; mahogany only), an inlaid cabinet model in similar style ("No.150," $150.00; dark mahogany, dull finish), and finally a ponderous horn-in-lid cabinet model with inlay and decorative veneering ("No.200," $200.00; dark mahogany, dull or polished finish). For an additional $50.00, the customer could have his Vitaphone equipped with an electric motor, wound for either direct or alternating current.

4-3. It was small internal-horn table models like this "No.25" which the Vitaphone Company intended to benefit from easy adaptability to external horns. By removing the coupling tube which connected the stationary soundbox with the neck of the internal horn, a metal flower or "Music Master" wooden horn could quickly be mounted. Little evidence of this practice exists, however, beyond references in catalogues. *Courtesy of the Domenic DiBernardo collection.* (Value code: G)

4-4. At $12.00, the Vitaphone Model "12" was the least expensive machine the company sold. A simple extension spring at the rear of the wooden arm exerted downward force on the record, in order to aid tracking and release as much physical energy as possible during play. All the elements of this machine have been reduced to make it affordable. *Courtesy of Sam Saccente.* (Value code: F)

4-5. The Vitaphone Model "100" ($100.00) made discreet use of inlaid wood. Note the articulated arm that connected the soundbox to the horn in the cabinet lid. This arrangement allowed the machine to be played with the lid open, a feature not available on the similar horn-in-lid machines being sold by Keen-O-Phone during the same period. *Courtesy of the Domenic DiBernardo collection.* (Value code: G)

4-6. A Vitaphone stock certificate, bearing the signature of Mr. Repp himself. *Courtesy of the collection of Howard Hazelcorn.*

4-7. The Vitaphone Company had a Canadian branch (in Toronto). One machine that seems to have been sold only in Canada is this "No.28." Like the Vitaphone floor models, the horn was contained in the lid of the cabinet. In order to create a connection between soundbox and horn on this particular model, the lid must be closed. *Courtesy of the Domenic DiBernardo collection.* (Value code: G)

4-8. A Vitaphone disc in its original sleeve, from the Canadian branch of the company. The discs were pressed by Columbia. Two machines in the Canadian line were depicted: on the left Type "No.28," on the right, Type "No.40."

4-9. A Vitaphone catalogue extolling the merits of C.B. Repp's reproducing system.

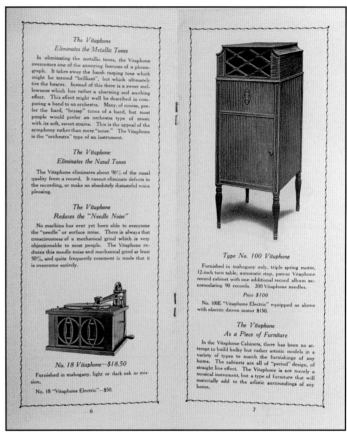

In April 1913, C. B. Repp, president of the Vitaphone Company, announced that he had devised a soundbox attachment which functioned as a "scratch absorber" for his machines. According to *Talking Machine World*, "It is used for the same purpose on these machines as one puts a filter on his water faucet....So important is this device considered that the production for three weeks was held up so that machines might be equipped." Although a thorough examination of numerous Vitaphone soundboxes has failed to reveal the nature of this "attachment," coincident with the introduction of the "scratch absorber," the company began exploiting the phrase "the perfect Vitaphone" in its advertising.

The rise and decline of the Vitaphone Company mirrored the trajectory of a similar independent, Keen-O-Phone of Philadelphia. Having worked vigorously for a few years to produce and market innovative disc talking machines, both firms faded away during 1914. An insight into Vitaphone's flagging fortunes may be inferred from a suit brought against the company in May 1914 by Fred Stern, president of the Newburgh, New York, chamber of commerce. Mr. Stern sought to recover $500.00 he had paid for Vitaphone stock based on the understanding that Vitaphone would locate a talking machine cabinet factory in his town. When a stock offering in Newburgh attracted insufficient response, Vitaphone had declined to continue the project. In truth, the Vitaphone Company was about to join the ranks of many other once-optimistic independents: trudging down the muddy road to oblivion.

4-10. The Keen-O-Phone "XXX" was the original incarnation of the Keen-O-Phone. Morris Keen, an entrepreneur and inventor working in Philadelphia, inaugurated his foray into disc talking machine manufacture with this statuesque model. The machine went through two distinct production variations, which mirrored changes in the Keen-O-Phone Company itself. The first version was known as the *"Keenolophone."* In a stroke of linguistic brilliance, Morris Keen incorporated both "phone" and "ola" into the name of his product, one-uping fellow independent manufacturers who usually appropriated one suffix or the other. *Courtesy of the Sanfilippo collection.* (Value code: E)

4-11. The "Keenolophone" epithet was eventually dropped, and the machine became known only by its numerical designation. Although the "XXX" (often referred to as the "30") was the most expensive Keen-O-Phone (introduced at $210.00 in 1911, soon rising to $225.00), an uncatalogued Model "30B" was also made. The external-horn machine pictured in this decal was actually produced in 1911, as the "Keen-O-Phone, Jr." or Model "I." A different external-horn model was also produced.

4-12. The interior of the 1911 "Keenolophone" displayed a number of interesting features. The horn was located in the cabinet lid, which had to be lowered to complete the sound passage. However, a door at the front of the playing compartment could be opened (as shown) to allow access to the interior controls. As the turntable rotated, it tracked beneath a stationary soundbox, which was adaptable to either lateral or vertical cut records. The mechanically-fed playing arrangement, though costly to produce, successfully discouraged accusations of infringement from Victor's legal department. The spiral sound arm was a Morris Keen invention (patent No. 907,814, December 29, 1908) described as follows: "By reason of this passage through this pipe...the sound is softened and the scratching and rasping tones...entirely obviated."

4-13. The graceful design of the Keen-O-Phone "XXX" record storage compartment suggested the soft curves of an odalisque. Shown is the second, 1913, version of the machine. In that year, the Keen-O-Phone Company reorganized in preparation of a huge marketing effort. Most Keen-O-Phone models were fitted with cost-reduced spring motors (the expiration of the "Berliner Patent" in 1912 had made mechanical-feed unnecessary), and the price of the "XXX" was reduced to $200.00 in September 1913 for the holiday season. More's the pity, Keen-O-Phone prices would drop even further when the unsuccessful company liquidated its inventory during 1914. *Courtesy of the Sanfilippo collection.*

Opposite page, top left: 4-16. A *New York Times* advertisement from December 1914 announced a half-price sale of Keen-O-Phones held at Gimbel's department stores in New York City and Philadelphia. The Keen-O-Phone Company had ceased business earlier in that year, and a large debt it owed to the Pooley Furniture Company (for cabinetry) was paid off in unsold talking machines. Through Gimbel's stores, Pooley as quickly as possible merchandised the hundreds of Keen-O-Phones it received. In the meantime, Pooley, crippled by the weight of this and other debt, was forced to declare bankruptcy. *Courtesy of Alan H. Mueller.*

4-14. This Keen-O-Phone "25" ("XXV"), at $125.00, occupied the third most expensive place in the 1911-1914 line, although the slightly dyslexic Keen-O-Phone numbering system listed the "20" ("XX," $175.00) above it. The "25" was the company's concession to the "high-legged" talking machine craze of the early 'teens which produced similar styling in machines such as Victor's first floor-standing Victrola "X," and Edison's Amberola "III." Like other top-heavy models of the period, the Keen-O-Phone "25" was meant to support a row of free-standing record storage albums on the low shelf. (Value code: F)

4-15. The motors in the larger Keen-O-Phones went through three distinct periods, which were reflected in changes to the motor board and positioning of the winding handle. The first era, 1911-early 1912, saw the use of a complicated and expensive mechanically-fed motor, which allowed the turntable to track under a stationery soundbox. In the second period, 1912 (shown here), the tracking system of the first motor was disabled to allow the machine to play as a conventional (non-mechanical-feed) talking machine. In the third era, 1913-1914, a smaller, much simplified motor was employed in most models, though the pilot holes for mounting the earlier motor were frequently visible.

4-17. A poster advertising the Alonzo Hatch Electro-Photo Musical Company's exhibition of moving pictures. These were not "talking pictures," so prominent mention was made of the pianist and tenor who accompanied the presentation. Contrary to popular impressions, "silent" films were never really "silent," because live performance was an integral and expected part of the entertainment.

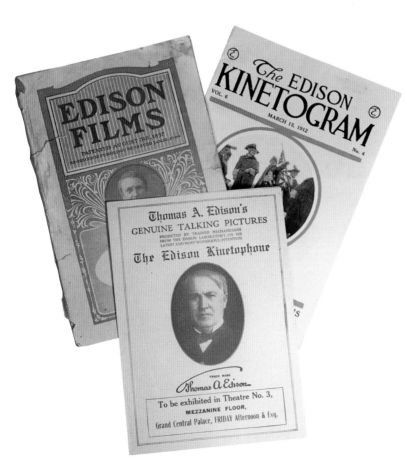

4-18. Clockwise from left: a 1906 catalogue of Edison Films, the March 15, 1912 issue of the *Edison Kinetogram*, a publication intended for Edison's moving picture exhibitors, and a flyer (5 3/4" x 9 1/2") advertising the exhibition of Edison talking pictures.

4-19. Edison's long experience with both movies and recorded sound naturally led him into the development of talking pictures. However, there were a number of significant hurdles to be overcome. Firstly, the sound had to be loud enough to be heard in a large theatre. Secondly, the sound and picture needed to be accurately synchronized. Thirdly, the records used to accompany the film had to be wear-resistant and unbreakable. By 1912, Edison felt he had achieved sufficient amplification (by licensing Daniel Higham's loud-speaking system) and synchronization. Unfortunately, Edison was still molding cylinder records from fragile "wax," since he lacked a patent to exploit durable celluloid. Furthermore, he did not consider the Diamond Disc records he was about to introduce suitable for sound-synchronized films. In 1912, Edison purchased the Petit celluloid cylinder record patent, and finally embarked upon the creation of an unbreakable cylinder for the Kinetophone, the Edison talking picture system exhibited 1913-1916. *Courtesy of the Edison National Historic Site.* (Value code: VR)

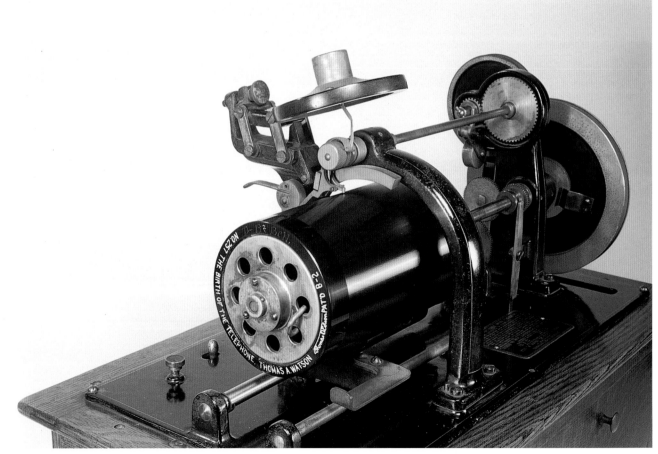

4-20. This close-up of the Kinetophone reproducing mechanism shows the elevated position of the Higham reproducer. The complete Kinetophone system consisted of a Phonograph and a movie projector (both electrically driven), an amplifying horn, film and records. In addition, other peripherals, such as a telephone hook-up between the Phonograph operator and the projectionist to fine-tune the synchronization, were necessary. *Courtesy of the Edison National Historic Site.*

4-21. The Edison Disc Model "A-60" was the least expensive member of Edison's initial 1912 disc machine line. In 1913, this oak "Mission" cabinet was substituted for an unsatisfactory one made of wood-grained metal. The $60.00 price of the machine, however, was still rather expensive for the average American. The Edison company was cutting itself a difficult (and ultimately unsuccessful) niche among the upper middle classes and the wealthy. *Courtesy of Lou Caruso.* (Value code: H)

4-22. The Edison Amberola "III" was introduced in August 1912 for $125.00. Equipped with the same mechanism as the "Opera" and the Amberola "IB," the Amberola "III" seems to have been the least popular of the three, judging by the number surviving. The shelf could accommodate three albums (which were separately available for $2.00 each) designed to hold a total of 90 cylinder records. The last Amberola "III"s were sold in mid-1915. *Courtesy of Jean-Paul Agnard.* (Value code: F)

4-23. The Autophone of 1913 was a simplified home-entertainment version of the Multiphone coin-operated cylinder apparatus. The carousel, for logistical purposes, was reduced in size to hold 12 four-minute records. The mechanism was powered, as in the larger version, by a stock Edison "Triton" motor. Rather than directing the sound upward, as in the Multiphone, the Autophone channeled the sound out through the lower part of the cabinet. The machine could be set to play or repeat in a number of configurations. Were extended excerpts from the classics ever issued on Blue Amberol cylinders, the Autophone would have been the perfect avenue for enjoying them. Unfortunately, Edison failed to aspire to more than independently sustained performances, and the complicated Autophone never rose above the level of an expensive novelty. *Courtesy of Jean-Paul Agnard.* (Value code: VR)

4-24. The Grafonola "Grand" in "Colonial" design was introduced in 1913 to complement an earlier version of the same machine with carved Queen Anne style legs, which had been released the previous year. The two "Grands" were promoted as the ultimate in Grafonola design. At $500.00, they included gold-plated hardware and tools, electric motor drive, and a set of leather-bound albums. Talking machine companies were fond of resuscitating older terminology. Hence, Columbia revived "Grand," and Edison re-used "Concert" and "Opera." In its closed position, the Grafonola "Grand" purposely gave few indications of its functional identity. The emphasis was on fitting into posh home décor. *Courtesy of the Domenic DiBernardo collection.* (Value code: E)

4-25. The Columbia Grafonola "Favorite" proved true to its name. It was a very popular model. When introduced in 1911, it looked like this, with two sound-modifying doors in front of the internal horn. Victor, which controlled the patent for that particular feature, was not amused. Columbia quickly substituted volume-controlling louvers that ironically would prove to be the most recognizable element of Grafonola design. (Value code: H, ordinary "Favorite")

4-26. This "Favorite" was factory equipped for electric operation. At a time when AC and DC power were battling for the America's homes, machines like this (and notably Victor's Electrolas) needed a formidable contraption such as seen here to accommodate either system. Note that the technology was borrowed from Columbia's Dictaphone division.

4-27. In April 1914, Edward N. Burns, vice-president of the Columbia Graphophone Company (as it was then being called) returned from a long trip abroad. He had visited Germany and contracted with a factory in Biersfield, Saxony, to produce a new line of Columbia disc talking machines. These inexpensively-made table models were named "Columbia-Europa," except in England where the "Regal" brand name was used. The machine shown here is the most commonly seen "Columbia-Europa" model. Following a formula already proven successful, the motor was small but the cabinet capacious enough to allow either an internal horn or an external bracket and horn to be fitted. Constructed of a pressed composition material, with metal trim, the hornless machine cost dealers only $3.00 and the horn machine $3.50, f.o.b. Germany. Needless to say, this enterprise was cut short by the outbreak of World War I. However, the "German connection" does explain why many external horn Columbias of the 'teens were supplied with the heavily fluted horns commonplace in Europe. *Courtesy of the Domenic DiBernardo collection.* (Value code: H)

4-28. The Columbia factory in Bridgeport, Connecticut.

4-29. A Columbia Graphophone Manufacturing Company stock certificate. *Courtesy of the collection of Howard Hazelcorn.*

Below: 4-31. This "Rex" table model displays definite links to Keen-O-Phone in various elements from the design of the soundbox to the shape of the crank. However, the "Rex," true to Mr. Wohlstetter's intent, was more cheaply constructed than anything produced by Keen-O-Phone. The Rex Corporation's sojourn in Philadelphia was to be brief. By April 1914, it had leased a building in Wilmington, Delaware, in order to relocate its manufacturing operation. Rex's record business would become the dominant aspect of the company, as the vertical record craze blazed during the middle-'teens. During the next three years Rex would press a large number of discs under its own name and some client brands, until it was purchased by the Imperial Phonograph Company in 1917.

Above: 4-30. A "Rex" talking machine, made by the Rex Talking Machine Corporation. In early February 1914, one of America's most innovative and well-constructed brands of disc talking machine was about to become defunct. Over the previous three and a half years, the Keen-O-Phone Company had been struggling to find adequate distribution for its distinctive roster of instruments. After the 1913 Christmas season failed to bring success, Keen-O-Phone decided to stop doing business. However, the fully equipped Keen-O-Phone factory on Orthodox Street in Philadelphia, and the firm's downtown vertical cut recording studio (the only "sapphire ball" record facility in the United States) were properties too valuable to squander. Early in 1914, the Rex Talking Machine Corporation was formed in Philadelphia with Philip Wohlstetter as president and H. W. Stoll as treasurer. According to an article in the February 1914 *Talking Machine World*: "...Rex...has leased the entire plant and equipment of the Keen-O-Phone Co. for a term of years, where it will manufacture a medium-priced line of talking machines and records...Thomas Kraemer is superintendent of the factory, while the recording end of the business is under the charge of Frederick W. Hager and Charles L. Hibbard." The key phrase in this quote seems to be "medium-priced." Mr. Wohlstetter himself referred to "...a line of machines and records of medium price, catering more particularly to people in moderate circumstances." Rex was not about to repeat its predecessor's mistake: "Keen-O-Phones" had been very substantially built, and rather expensive. (Value code: H)

4-32. Two "Rex" vertical cut 10" disc records. On the left, the earlier label, showing a king with a Rex talking machine. On the right, a record produced a year or two later, which resembled the look of the succeeding "Imperial" label.

The "Vertical" 'Teens and the Realignment of the Market Following the First World War

The proliferation of small disc talking machine brands in the 'teens was fed by the vertical cut recording system. Victor and Columbia patents controlled lateral cut recording, the standard of the industry. In the lateral disc, sound vibrations were stored on the walls of the groove. In the vertical disc, the vibrations were deposited at the bottom of the groove. The vertical system had proved successful in France, and it became the refuge of any American company which wanted to manufacture disc records during the 'teens. The sudden availability of vertical discs created an immediate need for machines equipped to play them. Victor and Columbia refused to make their talking machines adaptable to vertical records (although after-market converters were available). They had too big a stake in the propagation of the lateral system. It became a central purpose of the independent brands, therefore, to produce disc talking machines equipped with the kind of jointed tone arms that could easily convert between the rival formats.

In 1919, when a group of vertical cut record manufacturers fronted by Starr prevailed over Victor in patent court, the way was suddenly cleared for anyone to enter the lateral cut market. It was then that the death knell struck for the smaller vertical cut record companies who could not or would not embrace the lateral system. By 1920, the phonograph industry had undergone an abrupt transformation. The vertical cut disc was gone. Even Pathé, the brightest and the best supporter of the vertical system, defected. In 1920, Pathé inaugurated its "Actuelle" (up-to-date) lateral cut discs. Highly innovative "Actuelle" talking machines, intended to play both lateral and vertical records, would also be introduced, indicating the American branch's commitment to change. Pathé phased out sapphire ball discs, thereafter. Only the Edison company continued to support the vertical format with its colorless Diamond Disc line, to ever-declining popularity. Although the drastic end of the vertical record following the War caused a winnowing of small talking machine brands, a host of new independents was about to arise as the next decade dawned. The 1920s would prove an entirely different epoch in the history of recorded sound, and an unexpectedly glorious swan song for the acoustic talking machine.

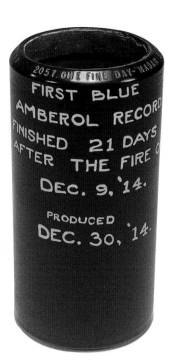

4-34. On the afternoon of December 9, 1914, the Edison factory in Orange, New Jersey, suffered a devastating fire. Although much of the Phonograph and record manufacturing facility was destroyed, the company rebuilt immediately. By December 30, Blue Amberol cylinder record production had been resumed. *Courtesy of the Edison National Historic Site.*

4-35. Few firms directly challenged Pathé's might in the French vertical cut disc market. One brand which did was "Radior," circa 1914. In this design, the horn may have receded into the cabinet, but the viewer is all the more aware of it for its oddly truncated appearance. The example shown has been fitted with an optional articulated adapter for playing either vertical or lateral records. (Value code: H)

4-37. The Pathé "Progress," when opened, shows the vertical-cut arrangement of the soundbox and the undisguised internal wooden horn. This machine was part of Pathé's British line, not offered in the United States. In the 1914 British catalogue this "Solid Oak Cabinet with Roll-top Cover" model was listed at £4 15 0. *Courtesy of the Domenic DiBernardo collection.*

4-36. The tambour top of the Pathé "Progress" was a clever alternative to the conventional lid, and lent an overall unusual appearance for a table model disc talking machine of the `teens. *Courtesy of the Domenic DiBernardo collection.* (Value code: H)

4-39. This Pathéphone floor-model, Model "7," was finished in the "golden oak" style which was rapidly being supplanted by mahogany in American homes. By the end of the 'teens, mahogany, both genuine and inexpensive "birch mahogany" finish, would dominate talking machine production. Note the articulated soundbox, the company's concession to convenience for the customer who wished to switch between vertical and lateral records. (Value Code: H)

4-38. The Pathéphone "25" was a popular alternative to the lesser Victrolas of the period, and was capable of playing both lateral and vertical cut discs. This was accomplished by the use of two separate soundboxes: a cumbersome system abandoned in later models that used a single "universal" soundbox. (Value code: I)

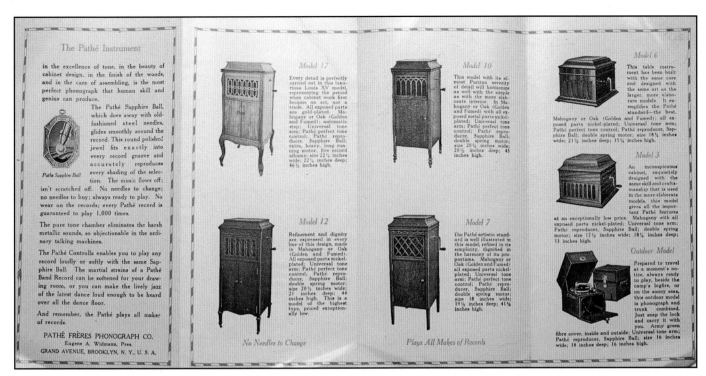

4-40. A fold-out flyer advertising the Pathéphone, from the World War I era. Note Model "7" is illustrated.

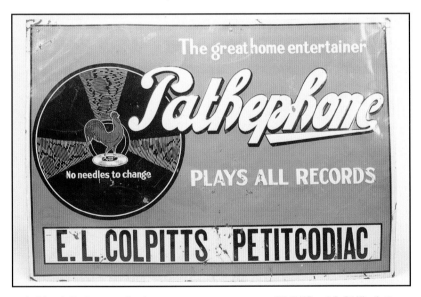

4-41. A Pathé metal advertising sign measuring 13 3/4" x 19 3/4", dating from the 1913-1920 North American sapphire disc period. To the usual "No Needles to Change" slogan have been added two lesser-seen mottoes. *Courtesy of the Domenic DiBernardo collection.*

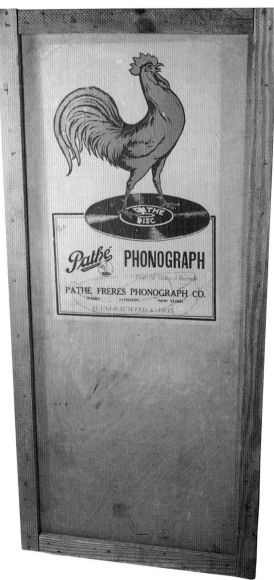

4-42. One side of a wooden shipping crate to enclose a Pathé upright phonograph. *Courtesy of Bob and Karyn Sitter.*

4-43. It is easy to mistake the "Baby" talking machine for a child's toy, but it was no such thing. In fact, The "Baby" was the only machine specifically advertised for use with "Little Wonder" discs (which weren't for kids, either). In the December 1915 issue of *Talking Machine World*, the "Baby" was offered for $3.95, "…playing three <u>full</u> LITTLE WONDERS with one winding." *Courtesy of the Domenic DiBernardo collection.* (Value code: I)

4-44. Unlike the "Baby," the "Baby Cabinet" was truer to its name. This was a miniature upright intended for children. *Courtesy of Sam Saccente.* (Value code: H)

4-45. From its lofty perch as purveyor of sublime and sophisticated music boxes at the turn of the twentieth century, the Regina Company descended by degrees. Having sold music box/Columbia Disc Graphophone combinations ("Reginaphones"), and cylinder playing coin-ops ("Hexaphones"), the firm began producing an inexpensive line of ordinary internal-horn disc talking machines in the 'teens. The "Reginaphone" name, with its classy associations, was continued. However, the most interesting feature of the low-end table model shown here was the unusual soundbox, similar to one used on a machine manufactured in Cleveland, called the "Bell." The prominent Regina insignia indicate the pride that lingered in a venerable company about to exhaust its place in the mechanical music field. *Courtesy of Steve and Ellie Saccente.* (Value code: H)

4-46. The ID plate of the "Reginaphone" table model shows the later-style "musical staff" logo that also appeared on the company's music boxes. The serial number beginning "402" fits into the numbering of the Regina "Corona" series sold in the late 'teens. By 1920, Regina had lost so much ground to rivals like Victor that it found its principal business to be vacuum cleaners and various other accessories. *Courtesy of Steve and Ellie Saccente.*

4-48. A Columbia dealer circular window decal, 15" in diameter, from the World War I era. *Courtesy of Martin F. Bryan.*

4-47. This large Grafonola was introduced in 1912 as the third model to sport the name "Deluxe." The size of this instrument can be realized by noting the 12" turntable. The four drop trays held a total of 60 records. Original price was $200.00. In 1915, this instrument's name was changed to the Model "200," in keeping with the policy of identifying models by their selling price, popular with many phonograph firms. Hardware, as befit the stature of this machine, was gold plated. *Courtesy of Lou Caruso.* (Value code: G)

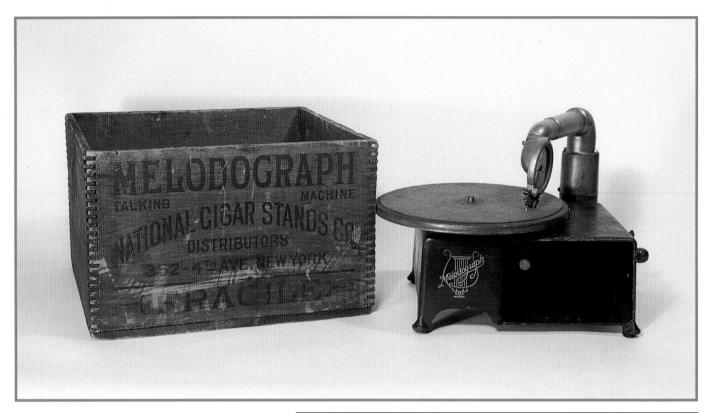

4-49. The Melodograph, sold by the Melodograph Corporation, New York, was a machine typical of the inexpensive cast iron brands that flourished during the 'teens. Its appearance was similar to the Garford/Vanophone of Elyria, Ohio, which was sold under client brand names. The Melodograph is shown next to its original shipping crate. From the markings on the crate, it likely that the machine was either offered as a premium or used in some other association with the sale of tobacco products. *Courtesy of Steve and Ellie Saccente.* (Value code: H)

4-50. This compact and curious-looking machine from the mid-'teens is often wrongly associated with "Little Wonder" records, since it was frequently sold under the identical name. In fact, there was no connection between "Little Wonder" lateral cut discs and this machine which played vertical cut discs. Many brands of small-diameter vertical cut records sprouted in the 'teens, creating a market for inexpensive machines such as this to play them. As may be seen from the decal, the "Little Wonder" machine was distributed under at least one other name: the "Vodaphone," sold by the "William Galloway Co., Waterloo, Iowa, U.S.A." (Value code: G)

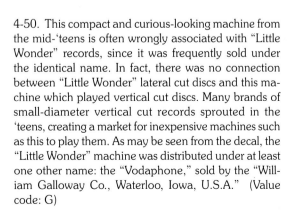

4-51. This inexpensive, pressed-steel talking machine was probably a premium for selling magazine subscriptions or flower seeds. Unlike similar instruments sold under trade names such as "Universal" or "Vanophone," this little machine carries no markings or clues to its origin. Despite the absence of a horn (the sound merely bounces around inside the steel "cabinet" and exits through the slots at left), the playing quality is surprisingly good. The two holes in front serve no purpose, but probably allowed the machine to be assembled using various motors. This particular type winds through the turntable, which was also common with children's phonographs such as those made by Bing. (Value code: I)

4-52. This Aeolian-Vocalion machine, circa 1916, was a posh model in Renaissance style, featuring wood burl and two built-in attachments: "Volunome" and "Graduola." The Aeolian company, pneumatic piano manufacturers, had introduced a line of disc talking machines in the mid-'teens. Aeolian was of the "intrusive" music lover school. The "Metrostyle" and later "Themodist" attachments on its piano-players and player pianos encouraged the operator to take an active role (pun intended!) in the mechanically-produced performance by manipulating hand controls. Likewise, the "Graduola" control was attached to the machine shown here via a cable which issued from the aperture at the left front, and permitted the listener to operate the "Graduola" from across the room, while the record played. As a 1916 Aeolian catalogue stated: "It will be noticed that the music is made gradually softer as the GRADUOLA knob is pulled out… Hold the GRADUOLA in the left hand so that the thumb may remain on top… Thus as the right hand lightly pulls or pushes the knob, the left thumb is guiding it so that all effects produced are gradual." One may imagine the pride of an "Aeolian" owner as he "helped" Mr. Caruso along with the aria! If the customer really wanted to feel like a professional, he could adjust the seemingly redundant "Volunome" to control the *dynamics* of the recorded performance. Enjoying records was hard work for the "Aeolian-Vocalian" owner. *Courtesy of Steve and Ellie Saccente.* (Value code: H)

4-53. An Aeolian Company stock certificate from the French branch of the company. *Courtesy of the collection of Howard Hazelcorn.*

4-54. An Aeolian-Vocalion advertisement showing an instrument in elegant surroundings. *Courtesy of Alan H. Mueller.*

4-55. In June 1915, Victor introduced the glamorous Victrola "XVIII" at $300.00 in mahogany. After 18 months, this expensive model was discontinued. During its short life, a customer could order the "XVIII" in Circassian walnut (shown) for $350.00. Not many such examples were manufactured. *Courtesy of the Sanfilippo collection.* (Value code: VR)

4-56. The Vernis-Martin cabinet, gold-leafed and expertly hand-painted in Louis XV style, was to offered by the Victor Talking Machine Company in 1910 for its Victrola "XVI." The "XVII" was a model introduced at $250.00 in December 1916 and for a short time made available in the Vernis-Martin finish (shown). The deluxe decoration added approximately $200 to the price of these instruments. *Courtesy of Robin and Joan Rolfs.* (Value code: VR)

4-57. A metal Victrola advertising sign, measuring 10" x 27 3/4".
Courtesy of Michael and Suzanne Raisman.

Victor Talking Machine Co.
 # VV-X- 271383J

4-58. The venerable Victrola "XVI" was the direct descendant of the very first Victrola model and continued to be a strong seller throughout the second decade of the twentieth century. The vast majority of these instruments were housed in mahogany cabinets. Shown is an exception: a Victrola "XVI" in Golden Oak. (Value code: H)

4-59. The Starr Piano Company of Richmond, Indiana, introduced a line of disc talking machines and records in 1916. The machines were well-constructed and incorporated interesting features. The Starr shown here was veneered in walnut. *Courtesy of William G. Meyer.* (Value code: H)

4-60. The nicely finished internal horn of the Starr, showing the pivoting metal plate by which volume was varied. Even this barely visible plate was given an "oxidized bronze" finish to match the rest of the hardware. *Courtesy of William G. Meyer.*

4-62. Kimball's take on the "His Master's Voice" trademark was a little girl listening to a Kimball, accompanied by the slogan "I want to see the lady <u>COME OUT!</u>" *Courtesy of Steve and Ellie Saccente.*

4-63. In acoustic talking machines, one of the few areas where minefields of patents did not restrict the imagination of designers was volume or "tone" modification. This Kimball used a very clever contrivance to control volume. A lever on the side of the cabinet activated a set of "shutters" on either side of the circular horn opening seen here. *Courtesy of Steve and Ellie Saccente.*

4-64. Manipulation of the control lever allowed the shutters to close over the mouth of the horn, variously restricting the sound passage. The completely muted position is shown here. *Courtesy of Steve and Ellie Saccente.*

4-61. Kimball was a Chicago piano manufacturer which took advantage of its woodworking facilities to produce a line of disc talking machines in the late 'teens. Although the design reflected obligatory elements directly derived from Victrolas, the heavy use of bright nickel in the playing mechanism lent an air of distinction. *Courtesy of Steve and Ellie Saccente.* (Value code: H)

4-65. A metal Kimball advertising sign which carries three slogans: the customary exclamation of the little girl, as well as "The TONE is so natural", and "Kimball Phonographs Play All Records." The advertising agency responsible for the Kimball account seems to have made a specialty of underlining for emphasis. Sign measures 11 3/4" x 35 1/2". *Courtesy of the Domenic DiBernardo collection.*

4-66. A record duster advertising a Kimball dealer.

4-67. The Cheney Style "Three" had the typically stout appearance of the firm's line of disc talking machines. Founded in 1914 by former classical violinist Forest Cheney, the Cheney Talking Machine Company used a sound passage and internal horn which were clearly designed not to infringe Victor patents. In spite of this, Victor sued, though the court eventually ruled in favor of Cheney. *Courtesy of Bob and Karyn Sitter.* (Value code: H)

4-68. Cheney concentrated on the construction of heavy floor-models. Even this table model conveyed the impression of weight and substance. *Courtesy of Bob and Karyn Sitter.* (Value code: H)

4-69. This extraordinary Cheney "Six" shows that the company was capable of style as well as substance. Cheney phonographs freely communicated their heavily-built nature through cabinetry which was interestingly-shaped but usually devoid of ornament. The expensive model shown, executed in elegantly figured oak, elevated the chubby body on gracefully sculpted Queen Anne legs. Note the "hanging" files for record storage, and the horn-covering grille hinged at the top as in most Cheney models. *Courtesy of the Sanfilippo collection.* (Value code: G)

Left: 4-70. The "Waddell" was made by the Music Table Company of Greenfield, Ohio, and oddly was referred to by the manufacturer as a "table" (No.MC858-R). Patented in 1916 and 1918, the playing compartment appeared ordinary, but the instrument held a clever secret. *Courtesy of Steve and Ellie Saccente.* (Value code: H)

Below: 4-72. The gimmick behind the "Waddell" was that after the record had been selected and the soundbox engaged, the internal horn was automatically uncovered as the operator closed the lid of the cabinet. This was certainly the best way of inducing the customer to play the record with the lid down, a practice that eliminated the extraneous noise caused by the steel needle. What appeared to be two "volume control" doors that would have infringed the Victor patent were, in fact, a single hatch cover, hinged at the top. The angle of this cover could be adjusted, or shut completely by pressing a button on the left side of the cabinet. *Courtesy of Steve and Ellie Saccente.*

Below: 4-71. A full view of the "Waddell" fails to reveal the secret. Record storage is the individual-divider type; the pattern of the machine, though boxy, appears conventional. Yet, one is tempted to ask how this small manufacturer could have flaunted the Victor patent on two sound-modifying doors over the mouth of the internal horn. Every other talking machine firm, including Columbia, had been forced to avoid the simple "opposed" door design. *Courtesy of Steve and Ellie Saccente.*

4-73. In a talking machine era where "cabinet was king," the Edison Diamond Disc "Army & Navy" Model had little to commend it. Indeed, its existence was due solely to the entry of the United States into World War I on April 6, 1917. By May 29, 1917, prototype models had been built, production orders went out on June 6, and the machine was publicly introduced on July 12. Intended to be taken to France by the American Expeditionary Forces, the "Army & Navy" Phonographs would provide soldiers and sailors with the civilizing, soothing influence of music. These rugged instruments were purchased by church and civic groups and given to local military units prior to departure. This model has the distinction of being the only acoustic talking machine designed specifically for military use in the First World War. Retailing at $55.00 with (supposedly) no profit for the company, the "Army & Navy" model was still far more expensive than smaller, lighter table model Victrolas such as the Victrola "VI" (which sold during this period for $25.00-$32.50). (Value code: G)

4-74. The interior of the Edison "Army & Navy" Model provided safe storage for reproducer, turntable and crank during transport over rough roads. Unlike other Diamond Disc models, the "Army & Navy" Phonograph included a crank escutcheon cover and a clamping device to secure the horn. In addition, each machine came equipped with an extra spring and barrel, graphite, grease and oil. Production of the "Army & Navy" Phonograph ceased in November 1918. Once back in the States, military outfits held lotteries to determine which servicemen would take the machines home with them.

4-75. A Victrola advertisement from 1918.
Courtesy of Alan H. Mueller.

4-76. A Victor catalogue from the First World War depicting doughboys waiting to be shipped out.

4-77. In this stately cabinet, the Edison company housed its Amberola "IB" cylinder Phonograph (introduced November 1911) and also its Diamond Disc "A-250" (introduced late 1912) and "B-250" (sold through 1916). Certain clues in this illustration, however, suggest that the machine shown here is none of the above. (Value code: VR)

4-78. In fact, this is not an Edison at all, but a Brunswick! Close examination reveals that a cabinet no doubt intended for Edison use has been factory-assembled as a Brunswick. It is well known that Edison engaged various outside manufacturers to supply the wooden cabinets for his Phonographs—sometimes on a catch-as-catch-can basis. It is also known that the Brunswick-Balke-Collender Company, which specialized in woodworking, constructed talking machine cabinets for client firms, including Edison. Clearly, a group of production "over runs" or cabinets refused by Edison were ultimately equipped and sold under Brunswick's own name. Brunswick introduced its own line of disc talking machines in 1916, the same year Edison was phasing out the Diamond Disc "B-250." This hybrid Brunswick, from the type of equipment it carried, including an early version of the "Ultona" soundbox, must have been assembled around 1918. This suggests that cabinets unwanted by Edison when the "B-250" was being discontinued had cluttered up a warehouse for a year or two before they finally reached the market as the Brunswick shown here.

4-80. What appears to be an Edison Diamond Disc "Sheraton Inlaid" model is actually a puzzle, the solution to which is found through phonographic "archeology." Edison did not make its own talking machine cabinets, choosing instead to contract with a number of furniture manufacturers throughout the eastern United States. For example, one of these cabinet suppliers was the Jamestown Mantel Company in Jamestown, New York. In September 1918, economic circumstances forced Edison to temporarily limit Phonograph manufacturing to only three models, which did not include the costly "B-275" ("Sheraton Inlaid") model. Evidently, one of Edison's cabinet manufacturers found itself with a cancelled order, and a number of these empty, expensive cabinets. In order to recoup its investment, the cabinet factory removed all labels and chalk markings normally seen inside these cabinets, equipped them with generic components, and sold the completed machines "out the back door." Note the non-Edison crank and the volume-control knob behind it. This particular example was found in the attic of a house in Jamestown, New York, suggesting that Jamestown Mantel might have produced it. (Value code: VR)

4-79. The Edison Diamond Disc "Sheraton Inlaid (plain)" was introduced in October 1913 under the designation "A-275." It initially sold for $275.00, years later increasing to $350.00. With minor changes, the machine enjoyed a long life in the Edison catalogue, finally disappearing in August 1927. The version illustrated is the "B-275," which was released in 1915. The machine resembled in general design the extremely popular "Sheraton" model "S-19," but was substantially larger and, of course, distinguished by its pretty inlay. (Value code: F)

4-81. Lifting the lid and removing t[he] grille reveals the true nature of t[he] "Jamestown Phonograph." Powered [by] a Heinemann No. "77" motor (Gene[ral] Phonograph Corporation, New Y[ork] City), equipped with a "Mutual" to[ne] arm (William Phillips Service, New Y[ork] City) and employing additional comp[o]nents of the Universal Stamping [&] Manufacturing Company of Chicag[o,] this machine can be dated to the l[ate] 'teens. There is no trace of Edison a[n]cestry within the cabinet, such as ho[les] for Edison motor braces or the ho[le] support post distinctive of Edison D[ia]mond Disc Phonographs. Instead, the[re] is a simple, neatly constructed wood[en] horn resting on two small legs. Su[ch] surreptitious phonograph "manufact[ur]ing" using Edison cabinets was carri[ed] out on a very small scale, and probab[ly] only under very special exigencies.

4-82. Another use of the name "Autophone," this time a short-lived talking machine introduced in 1919 by the Autophone Company of 117 Cypress Avenue, New York City. The machine is shown in position to play vertically cut records, but the soundbox could be removed and re-inserted in the tone arm to play laterally cut discs as well. The odd arrangement of the "Autophone's" tone arm, crank and horn made it rather cumbersome to use. The "Autophone" name appeared on at least two other unrelated and vastly different types of talking machine. (Value code: I)

4-83. The Puritan was a late 'teens disc talking machine with a rather matronly profile. The designer must have reasoned that locating the horn opening at the bottom of the cabinet would allow the sound passage to achieve considerable length and, therefore, greater volume. As a 1919 advertisement stated: "...the LONG WOODEN HORN... is an exclusive, patented feature, AND CANNOT BE USED BY ANY OTHER MANUFACTURER." Unfortunately, the merit of the reproduction was better appreciated by persons reclining on the floor. Unless one were planning a Roman dinner party, the dog and the cat had the best seat in the house. *Courtesy of Robin and Joan Rolfs.* (Value code: G)

4-84. The interior of the Puritan displayed equipment typical of the period, including a vertical record adaptable soundbox. The machine was made by the United Phonographs Corporation of Sheboygan, Wisconsin. *Courtesy of Robin and Joan Rolfs.*

4-86. A Brunswick advertisement showing one of the company's deluxe "console" models. *Courtesy of Alan H. Mueller.*

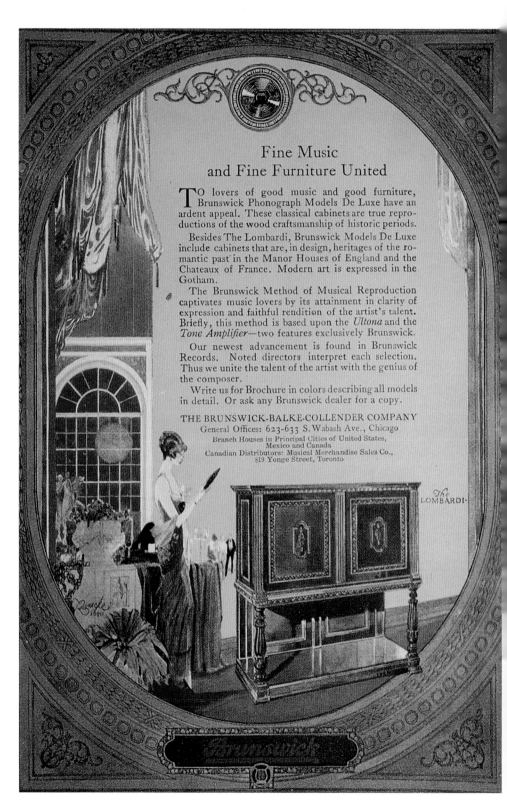

4-85. The Brunswick-Balke-Collender Company overtook Edison in 1921 as the third largest talking machine manufacturer, behind Victor and Columbia. This Brunswick Model "135" in walnut was representative of the reason for the company's success. In addition to the attractive cabinet, the machine was equipped with Brunswick's "Ultona" soundbox. This device, patented September 18, 1917, enabled the machine to play three types of disc records: lateral, vertically cut sapphire ball, and vertically cut Edison records. *Courtesy of the Sanfilippo collection.* (Value code: F)

THE CABINET IS KING

The end of World War I brought a tremendous wave of prosperity to the United States. Household goods, including talking machines, enjoyed strong sales in the 1919-1920 period. Seemingly overnight, dozens of talking machine manufacturers took root and began offering a plethora of brands from which to choose. Most of these new phonographs featured motors and hardware offered by a relative handful of suppliers. Otto Heineman's Phonograph Supply Company was a prominent purveyor of "generic" talking machine parts including motors, turntables, tone arms and soundboxes. Other concerns, such as the Fletcher-Wickes Company and the William Phillips Service, specialized in tone arms. Still others supplied simple parts such as crank escutcheons, knobs, speed control plates, and customized nameplates. Consequently, several dozen phonograph brands might share the same motor, scores would be equipped with the same tone arm, and much of the attendant hardware might be identical. Usually, the only distinguishing feature of these "off-brand" machines was the cabinet. Except for big names such as Victor, Columbia, Brunswick, Edison, Sonora, Pathé, and Cheney, the American talking machine industry took on the appearance of a huge family reunion: each built pretty much the same, but dressed differently.

5-1. Cleverly concealed in this mahogany secretary is an unknown make talking machine of European origin. The green cloth at top covers the mouth of the horn. Not shown is a wooden cover which, when in place, conceals the turntable and tone arm. The sound box is marked: "Prima Starkton Concert." Efforts to disguise phonographs as furniture were popular during the 'teens and early twenties. 33 1/4" wide x 20 1/2" deep x 73 3/4" tall. *Courtesy of the Sanfilippo collection.* (Value code: VR)

5-2. The Heywood-Wakefield Company of Boston, Massachusetts, marketed a line of talking machines in wicker cabinets which they called "Perfek'tone." The soundbox was sheathed in rubber, and the horn was "...composed of a matrix of wood and fabric having a peculiar vibratory action of its own, and gives a fullness and sweetness of tone which can be compared to a rare old violin." The main selling point was obviously the wicker cabinet, and the "Perfek'tone" catalogue pulled out all the stops in extolling its superiority. "The Perfek'tone Cabinet is the last word in acoustical science as applied to sound-reproducing instruments, having no confined air spaces or cavities to destroy the original coloring of the music. The counter vibrations, so noticeable with wood cabinets, are entirely eliminated by the use of reed and cane." *Courtesy of Robin and Joan Rolfs.* (Value code: G)

5-3. The interior of the "Perfek'tone." This example was finished in "Old Ivory," but several other colors were available including "Verd Mahogany," "Holland Gray," and "French Walnut." The advertising hype could not disguise the fact that the "Perfek'tone," like many others, was forced by the Victor patent to use a non-tapered tone arm, which compromised sound quality. *Courtesy of Robin and Joan Rolfs.*

5-4. The Arion Manufacturing Company of Boston, Massachusetts, is best remembered for producing a compact, lidless table model disc talking machine. However, during the late 'teens, the company also sold larger and, in this instance, luxuriously-appointed machines. The signature feature of the Arionola was a "reflex" horn. Behind the decorative grille in this illustration is a concave wooden shell. The vibrations from the soundbox were directed at this enclosure and reflected out toward the listener. *Courtesy of Sam Saccente.* (Value code: H)

5-5. The Arionola tone arm was located in a forward position, unusual for acoustic disc talking machines. The use of a reflex horn made this arrangement possible. *Courtesy of Sam Saccente.*

5-6. The cabinet of this Arionola was artfully inlaid. Very few American talking machines were decorated in this Old World style. *Courtesy of Sam Saccente.*

This emphasis on cabinetry was a boon to many furniture factories, whose cabinet designs distinguished each phonograph brand. Some furniture makers offered empty cabinets to the trade to be fitted out and labeled by any "manufacturer" who wished to jump onto the bandwagon. For many years, Grand Rapids, Michigan had been a center for furniture building. By 1919, it became a hub for talking machine cabinet manufacture, with numerous firms vying for the "off-brand" business. The more prestigious firms built cabinets for only one client, as the Herzog Art Furniture Company (of Saginaw, Michigan) did for the Sonora Phonograph Company. As Victor, Edison and Columbia had earlier discovered, the buying public was primarily interested in the cabinet, not the mechanism.

This was a noteworthy shift in talking machine marketing. At the turn of the twentieth century, the major talking machine companies had been offering different mechanisms in virtually identical cabinets. Price variations among different models were due primarily to the power of the motor, list of features, or size of the horn. Victor had offered a distinct range of cabinets in its "I" through "VI" series beginning in 1904. By August 1906, the Victor "VI" mechanism was available in a floor-stand-ing cabinet called the "Victrola." Columbia quickly grasped the significance of the "cabinet" development, and offered its first internal-horn machine (the "Symphony Grand," with a cabinet disguised as an upright piano) in March 1907. Columbia would coin the name "Grafonola" to distinguish its cabinet models. Edison finally introduced a Victrola-like cylinder Phonograph (the "Amberola") in late 1909, but with its own unique playing mechanism.

Not until the advent of the Diamond Disc Phonograph in 1912-1913 did Edison offer a true choice of cabinet styles. By that time, all three of the major talking machine companies were well on the way toward standardizing the mechanics of their respective products, focusing creative efforts on cabinetry. The "off-brand" phenomenon of the 1916-1921 period was the culmination of this trend: dozens of different companies employing the same mechanical components to equip their own cabinets. Over the previous fifteen years, the internal workings of the talking machine had become a matter of little consequence to the buying public. The visual aesthetics of cabinetry had overshadowed the miracle of a machine that talked.

5-8. By the time this Sonora phonograph was manufactured around 1920, public taste in the United States had embraced mahogany as the wood of choice for home furnishings. During the first decade of the twentieth century, oak had been enormously popular, and most talking machines had been made of it. In the 'teens, oak began to be viewed as *déclassé*, and an increasing number of talking machines, especially internal-horn models, were constructed of mahogany. It is a great treat to find this Sonora "Baby Grand" in golden oak. *Courtesy of Dan and Sandy Krygier.* (Value code: G)

5-9. The Sonora "Elite" was a middle-range machine in a line which emphasized high-class, tony appeal. The 1920 catalogue reads like a social register of talking machines, self-consciously describing the wooden soundbox and tone arm as follows: "Being equipped with a Sonora All Wooden Tone Passage, it embodies every refinement in tone reproduction that can be desired." Indeed, the pleasingly pneumatic design of the cabinet probably convinced most customers. *Courtesy of Dan and Sandy Krygier.* (Value code: G)

5-10. This plump Sonora "Portable," circa 1920, covered in natural leather, incorporated a reflex sound dish in the lid. The catalogue described it as follows: "The Sonora Portable is where you want it—when you want it—without a bit of trouble or inconvenience…" (Value code: H)

5-11. This 9" diameter plaster wall plaque identified Sonora dealers. Although the plaque indicated that the company had been founded in 1914, the Sonora brand name appeared five years earlier. By 1914, however, the original firm had effectively ended, and a completely reorganized Sonora was established. *Courtesy of the Charles Hummel collections.*

5-12. In the Columbia catalogue (1920), this outfit was identified as the "Type D Grafonola and Pushmobile." It was intended for use in schools, where the rubber-tired wheels would facilitate its mobility. The example shown is in oak, but mahogany was also available. The identical wheeled cabinet had been advertised for use with no particular brand of machine in November 1915 by Schloss Brothers of New York City. "It gives you the advantage of taking both machine and records from room to room wherever and whenever the occasion requires. Particularly suitable for schools." *Courtesy of the Domenic DiBernardo collection.* (Value code: G)

5-13. The Grafonola school outfit was introduced in 1918 for $125.00. Institutions could purchase the outfit for the discounted price of $95.00. Interestingly, the 1920 Grafonola catalogue describing the "Type D Grafonola and Pushmobile" stated: "We no longer manufacture the outer-horn-type instruments as they are considered obsolete and are not desired by the general public, because of their unsymmetrical appearance, and also because of their greater liability to damage." Noticeably absent from this explanation was any mention of comparative sound quality. The "obsolete…outer-horn-type instruments" could still outperform the new "Type D Grafonola" acoustically, despite its touted aesthetics. *Courtesy of the Domenic DiBernardo collection.*

5-14. A double-sided porcelain Columbia dealer's sign, 28" diameter. *Courtesy of the Domenic DiBernardo collection.*

5-15. The Columbia Grafonola Model "L-2" was introduced in 1918 for $225.00. By 1920, the "L-2," with its distinctive lid and Sheraton style cabinet, was being prominently featured in Columbia advertising. The price of the "L-2" rose to $275.00 in 1920, but in 1921 (in a manner now typical of Columbia) was dropped to $175.00. This model was available in either American walnut (shown), or English brown mahogany. *Courtesy of William G. Meyer.* (Value code: H)

5-16. A Columbia metal sign measuring 17 5/8" x 23 5/8". *Courtesy of Michael and Suzanne Raisman.*

5-17. A Columbia advertisement from December 1920. *Courtesy of Alan H. Mueller.*

By late 1920, inflation had begun to decay the post-war American prosperity. The price of virtually everything, including the talking machine, began to rise and kept rising. This resulted in an economic recession in 1921. That year, Columbia's sales plunged alarmingly. This long-established company managed to hold out until October 1923, when it went into receivership, but for the hundreds of fly-by-night talking machine companies that had sprouted in the postwar period, death was instantaneous.

Victor, Brunswick and Edison all suffered, but did not share Columbia's fate. Victor managed to show slight sales increases during 1921-1923, but a new force combined with difficult economic times would drive Victor's sales down considerably by 1925. This new threat was radio.

A Menace in the Air

Regular radio broadcasts began in October 1920. The American public embraced radio with all the enthusiasm of a child on Christmas morning. Sales of crude radio receivers skyrocketed. Fledgling radio manufacturers sprang up as rapidly as phonograph firms had after the war. People were fascinated by the new toy. At first, reception was poor, synchronizing the settings on early radio sets was tricky, and only one person at a time could listen through a headset (much like the listening tubes necessary on the earliest talking machines). Yet, the thrill of picking up a signal from hundreds of miles

away was inescapable. Radio listeners began keeping "logs" of stations they had picked up on clear nights. Radio was irresistible to a world that valued speed and being "up-to-date." Not only was the listening "free," but radio offered an immediacy that starkly contrasted the talking machine's ability to encapsulate and preserve.

Ironically, radio development closely mirrored that of the talking machine. Perhaps the most similar pattern was the rapid evolution from earphones to horn-type speakers to apparatus concealed in cabinets. Attendant improvements in the receivers themselves greatly enhanced fidelity and ease of operation. All this activity siphoned off a significant amount of potential talking machine business. The Radio Corporation of America reported a profit of $26,394,790 in 1923, far more than the gross sales of most talking machine companies. In 1924, while Victor's total sales were $36,951,879, RCA's *profits* were a staggering $54,848,131. Both Brunswick and Sonora began selling radio/phonograph combinations, but the limitations of current acoustic recording and reproducing systems were becoming painfully apparent.

The talking machine industry of the early 1920s was offering products mired in decade-old styling and technology. Although some of the "off-brand" manufacturers had marketed a few forward-thinking cabinet designs, they were hamstrung by the usually mediocre mechanical and acoustic components available to them. If advances in the talking machine art were to be made, it would be up to the giant corporations to make them.

5-18. The Edison cylinder Amberola line consisted of only three models for most of its life. After 1915, the "30," "50," and "75" were the only choices Edison offered to the cylinder-buying public, except for the briefly-appearing "60" and "80." Shown is a Babson company (mail order) catalogue illustrating the Amberola "75," rather plain for a "flagship" model. An Amberol record supplement is shown at upper left.

5-19. Edison Diamond Disc Phonographs were expensive and less portable than Victrolas or Grafonolas. In February 1919, Edison introduced "The Bungalow Model" (relatively inexpensive, for an Edison, at $95.00) to be used as a portable "camp" instrument. The model name was changed to "Chalet" in April, but the identification plates always read "B-19." This was the only Edison cabinet made of red gum, an inexpensive wood used in many home interiors of the period. *Courtesy of Lou Caruso.* (Value code: H)

5-20. An 11 3/4" x 22" trolley sign from the late 'teens advertising the Edison Diamond Disc Phonograph. *Courtesy of the collection of Howard Hazelcorn.*

5-21. An 11 3/4" x 22" trolley sign promoting the Edison "Tone Tests." The tests were public demonstrations of live versus recorded performances. Beginning in 1915, the Edison company held these highly publicized events all over the country. *Courtesy of the collection of Howard Hazelcorn.*

5-22. Ediphones evolved into enclosed metal floor units, painted gray or drab green. Acoustic recording and replay of cylinder records remained remarkably unchanged from the technology of the1890s. Whereas this appears to be a typical later Edison dictating machine, it is actually a mere 6 1/2" high! This convincing miniature was used in Edison marketing. *Courtesy of the Charles Hummel collections.*

We have seen that Columbia was preoccupied with its own grave financial situation, and was not in a position to embark on expensive research and development projects. Although by 1921 Brunswick was the third-largest phonograph producer in the United States, it was nevertheless prevented by key Victor patents (such as that for the tapering tone arm) from offering first-class sound reproduction. Brunswick countered these disadvantages by developing a soundbox with a very wide diaphragm (the "Ultona"). Brunswick cabinets, however, were beautifully designed and well built. Brunswick prospered in the early 1920s primarily through the quality of its cabinetry. Edison, by contrast, was cheapening its

cabinets throughout this period. Although the company was a pioneer in offering "Period Cabinets" based upon recognized furniture styles, these elaborately designed Edison Disc Phonographs were at first exorbitantly priced. By the time Thomas Edison himself decreed that all Edison Phonographs would be housed in "Period Cabinets" (albeit much simpler designs), other talking machine companies had surpassed Edison in cabinet quality. This left the Edison company relying on its aging vertical-cut mechanics and rather stodgy recorded repertoire: both liabilities in the fast moving 1920s. Victor alone seemed in a position to advance the talking machine art.

5-23. In 1920, the Victor Talking Machine Company introduced a limited series of "period style" Victrolas in "wide" cabinets. These expensive Victrolas (starting at $500.00) represented such schools of design as Chippendale, Hepplewhite and Queen Anne (shown). As the first examples of Victor console models, these "wide" Victrolas all shared the same curiously *stacked* appearance. The richness of the veneer, however, made this particular piece undeniably stunning. *Courtesy of the Sanfilippo collection.* (Value code: VR)

5-24. The rare Victrola "XIII" (1921) presents something of a mystery. It appeared to be a simplified version of the first style Victrola "XIV" of 10 years earlier. However, the price ($250.00) must have seemed excessive for such a plain-looking model. According to Victor documents, a mere 662 VV-XIIIs were manufactured. Curiously, the example shown is numbered 863, indicating that the numbering probably did not start at "0." *Courtesy of the Domenic DiBernardo collection.* (Value code: VR)

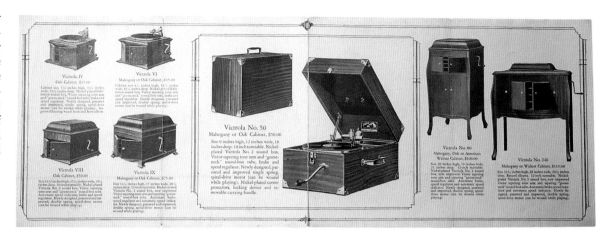

5-25. A fold-out flyer from May 1922, featuring a range of Victor machines including the Victrola "50," the company's first "portable," and (far right) the "VV240," one of the unsuccessful and derisively-called "humpback" models.

Today in America, we tend to think in "eras": "Seventies Disco"; 1930s-Depression Era; "The Roaring Twenties." What is missing in these tags is the understanding of how historical epochs blended into one another, with embers of the previous period flickering out during the next. Although it is conventionally believed that external-horn talking machines disappeared during the early 'teens, and that Victor was the first to discontinue them, the instrument shown here proves otherwise. This mahogany Victor "V" was made in the early 1920s, and represented the very last incarnation of the Victor (as opposed to "Victrola") talking machine. Early in the 'teens Victor had discontinued the costly-to-build mahogany Victor "VI" and designated the oak "V" the company's premier external-horn machine. Over the next decade, the "V" absorbed various Victrola elements: motor, turntable, brake assembly, tone arm, etc. Finally, the "V" was completely redesigned in the early 1920s and recast as a curious ghost of the long dead "VI." Observe that the cabinet is mahogany, but also that it is *huge*, as can be seen from the relative size of the 12" turntable. The motor is a typical four-spring Victor mechanism such as the upper-range Victrolas employed (note the speed control window at the right front). There is a Victrola automatic brake, an inverted Victrola-type tone arm and a Victrola "No.2" soundbox. The mahogany horn is identical to the smooth (oak) horns used on Victrola "XXV"s (the school machine). Many of these talking machine "dinosaurs" were exported by Victor to countries where external-horn instruments were still commonplace. Some were sold to American audiophiles, to whom this machine, with its up-to-date mechanics and superior external-horn sound, would have represented the height of listening pleasure. (Value code: VR)

5-27. A special genre of portable acoustic disc talking machines grew up, beginning in the early 1920s, around the rather peculiar premise of disguising them as box cameras. An incredible outpouring of imagination was funneled into this concept. Every effort was made to make these instruments indistinguishable from cameras, yet fully functional as talking machines. Shown here is the aptly named German "Kameraphone" from the twenties. The little leatherette-covered box opened to reveal all the parts stored within. The turntable consisted of spokes on which a regular 10" 78 rested. The soundbox was an unabashed copy of the Victor "Exhibition" (marked "Exposition") that directed the sound into a celluloid resonator in tortoise-shell finish. *Courtesy of Walter and Carol Myers.* (Value code: H)

5-29. Here, the "Peter Pan" is shown with all its parts carefully compacted for travel. The claim "MADE IN ENGLAND, Swiss motor" might better be interpreted "assembled in England." The component parts of the "Peter Pan" had a decidedly Continental look about them. *Courtesy of Walter and Carol Myers.*

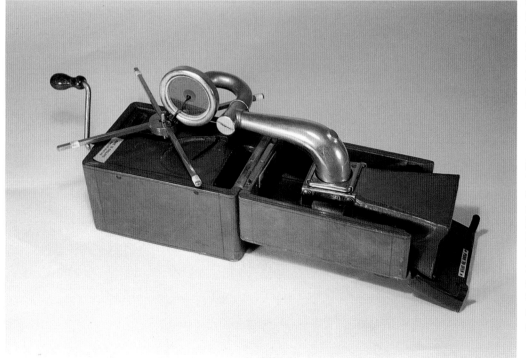

5-28. The "Peter Pan" was a popular "cameraphone" offered in Great Britain beginning in the 1920s. The better models sold under this brand were leather-covered and featured bellows-like leather horns. The spoked turntable was also employed by other marks, but the "Peter Pan" came closest to a conventional tapered tone arm with Victor-inspired U-joint. *Courtesy of Walter and Carol Myers.* (Value code: H)

5-30. This Peter Pan model employed a telescoping horn, which conveniently collapsed to fit inside the little case. Woe to the picnicker who lost a single segment of this clever horn! (Value code: H)

5-31. What could be more "cameraphone-like" than a "Cameraphone?" The brand name of this model left nothing to the imagination. Again, a spoked turntable has been employed, but this particular version disassembles rather than folds. One spoke has been removed to illustrate the principle. *Courtesy of Walter and Carol Myers.* (Value code: H)

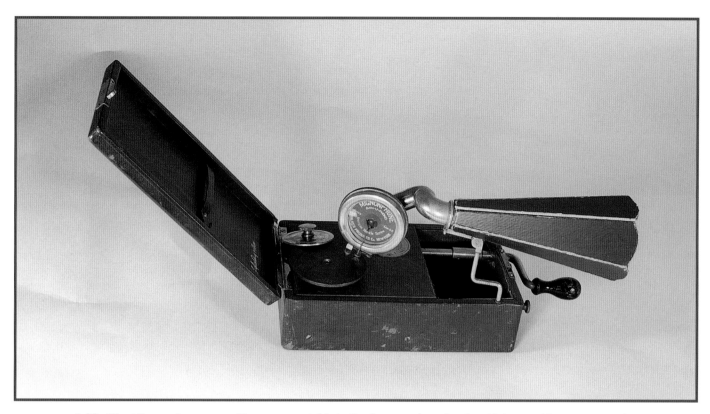

5-32. The Mignonphone was a European portable in the "cameraphone" style with box and horn customarily covered in leather. This particular example used a folding cardboard horn. It was available in various colors that included tan, red, blue and green. A key feature was the collapsible horn, which was larger than most "cameraphones" could accommodate. This model was sold by L.S. & E.H. Walker, New York, who directed the client to cold-bloodedly murder French pronunciation by saying "Min-u-phone." *Courtesy of Walter and Carol Myers.* (Value code: G)

THE SPARK OF INNOVATION

In early 1924, the Bell Telephone Laboratories and Western Electric arranged a special demonstration for officials of the Victor Talking Machine Company. Western Electric engineers, under the supervision of Joseph P. Maxwell, had developed an electromagnetic recording head that was capable of making high-quality electrical recordings. Additionally, Henry C. Harrison had designed an acoustic talking machine that incorporated a horn of precise relative length, rate of taper, and size of opening. Known as an "exponential" horn, this apparatus, coupled with an improved soundbox and tone arm, gave vastly superior performance. When playing the new electrically-recorded discs, the results were truly impressive. Victor had its own people working on a method to record electrically, and was confident that a comparable acoustic playback system was possible. Victor decided to put off Western Electric's offer to license the system, in favor of developing its own product. It was a dangerous gambit.

While Victor vacillated, word of the Western Electric recording process leaked out. The Bell Laboratories' electrically-recorded masters were processed by the Pathé record pressing plant in Brooklyn, New York. In Decem-ber 1924, industry veteran Frank Capps was visiting the plant, and heard some of the new electrically-recorded discs. He immediately sent some extra copies to Lewis Sterling in London, who was in charge of the English branch of Columbia. Sterling listened to the records and, fully awake to their significance, "caught the next boat" for New York. There he met with representatives of Western Electric and the newly reorganized Columbia Phonograph Company. Because Western Electric would not license a foreign company, Sterling purchased the American Columbia Phonograph Company for $2,500,000. In February 1925, he obtained a license from Western Electric to exploit the new electrical recording/acoustic reproducing method through Columbia. Sterling's English Columbia was licensed as an affiliate of the American firm.

Seemingly in the nick of time, Victor asked for another demonstration of the Western Electric system in February 1925, and obtained a license in March, a few weeks after Columbia. In the late spring of 1925, both companies began to release electrically-recorded discs quietly, so that remaining stocks of acoustically-recorded records could be sold. Nothing on the labels of the new discs revealed that they had been recorded with the revolutionary new process. A sharp-eyed Victor consumer might have noticed, just outside the label area near the run-out groove, a small oval with "VE" stamped inside. This was the only clue to the true nature of the recording ("Victor Electric"). For the moment, the discs' improved sound went largely unnoticed as played on obsolescent Victrolas and Grafonolas. Residual stocks of these machines and acoustically-recorded records were cleared out during the summer and early fall of 1925.

November 2, 1925 was heavily advertised as "Victor Day," when the new "Orthophonic" Victrola, equipped with an "exponential" horn, was introduced. Columbia unveiled its line of exponential horn machines in January 1926, referring to them as "Viva-tonal" some months later. Now the full impact of the new electrically-

recorded records could be demonstrated. The first major improvement of sound reproduction in decades was finally at hand. Perhaps even more noteworthy, and chilling for Victor and Columbia, was the fact that Brunswick had made a similar achievement. In November 1925, Brunswick introduced its "Panatrope," along with its own electrically-recorded discs. Unlike the talking machines of Victor or Columbia, the Brunswick "Panatrope" was completely electric, and included a radio. The Brunswick electric "Light Ray" discs were recorded through a system developed by General Electric, originally intended for use in recording motion picture soundtracks directly on film. Early 1926 marked a new beginning for the talking machine industry.

5-33. The Brunswick "Cortez" imitated the mid-1920s styling made popular by Victor's "Orthophonic" Victrola line. *Courtesy of Dan and Sandy Krygier.* (Value code: H)

5-34. The interior of the Brunswick "Cortez." Although features such as tapered, off-set tone arms, more acoustically efficient horns and scientifically designed soundboxes were often-copied Victor innovations, they undeniably elevated the standards of the entire talking machine industry. *Courtesy of Dan and Sandy Krygier.*

5-35. The Orthophonic Victrola "8-9" was introduced in 1928 at $175.00. It was a considerable departure for the Victor Company. Though it was based on the "8-8" school Victrola, which had minor blue-painted accents on the cabinet, the "8-9" was brightly, almost gaudily, decorated in turquoise blue, gold and red (on the doorknob escutcheons). No other Victor machine could be said to be so vividly enhanced. *Courtesy of Michael and Suzanne Raisman.* (Value code: G) **(see next page for interior view)**

5-36. The interior grille of the "8-9" was brilliantly highlighted. One wonders how many patrons could have fit this beaming fellow into their home décors. *Courtesy of Michael and Suzanne Raisman.*

Opposite page:
Top: 5-37. A Victor "Orthophonic" catalogue illustrating on the left the Victrola "10-50" with automatic record changer. On the right is the Victrola that rewrote the standards for the talking machine industry: the "8-30," first known as the "Credenza." Although it was large, it lacked significant record storage, since most of its bulk was given up to a formidable exponential horn. The wholehearted acceptance of the "8-30" signaled an end to the hegemony of the talking machine cabinet. The quality of the sound was once again of primary importance. *Courtesy of Alan H. Mueller.*

Bottom: 5-38. The "Orthophonic" catalogue listed everything from electric behemoths that changed their own records to the humblest table models and portables. *Courtesy of Alan H. Mueller.*

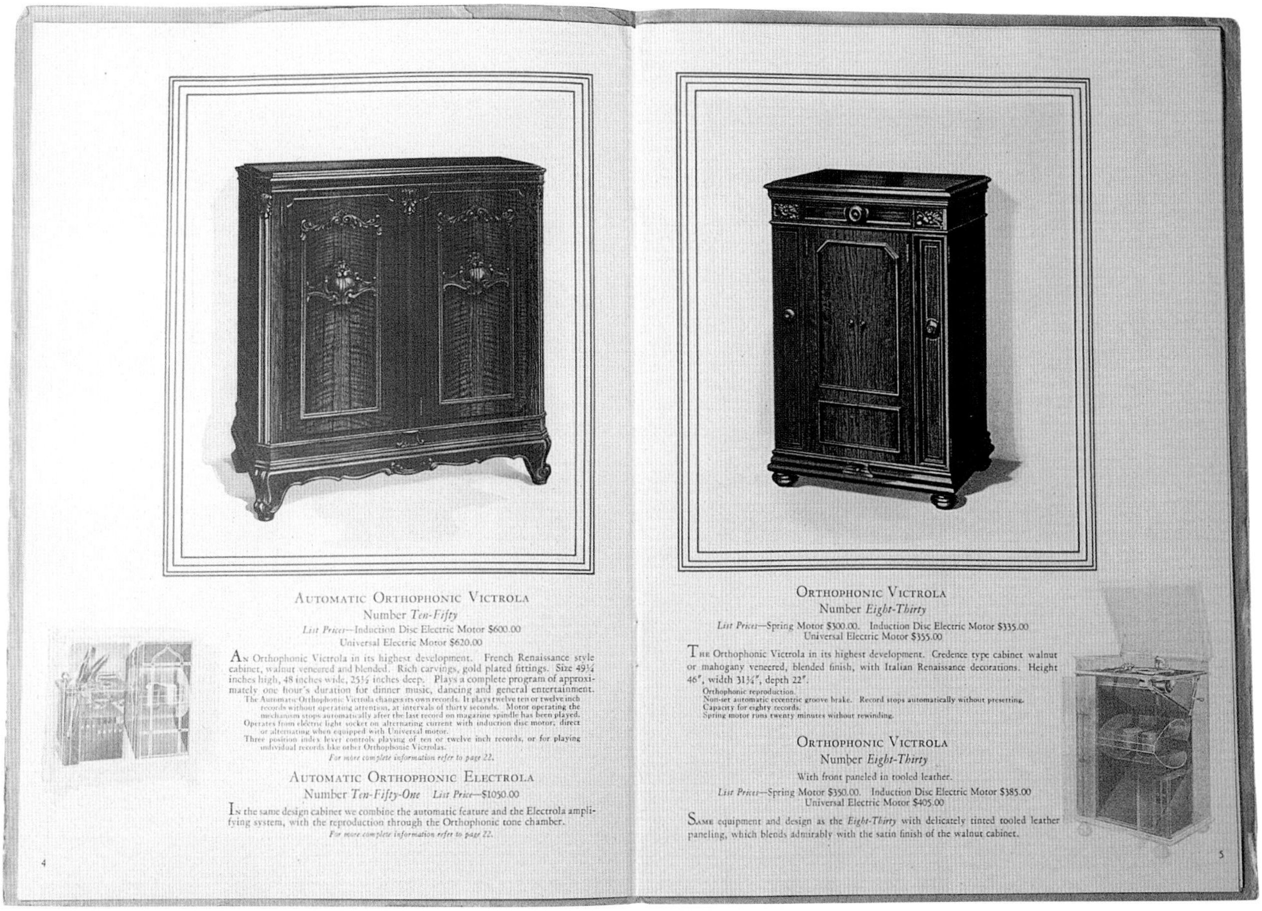

AUTOMATIC ORTHOPHONIC VICTROLA
Number *Ten-Fifty*

List Price—Induction Disc Electric Motor $600.00
Universal Electric Motor $620.00

An Orthophonic Victrola in its highest development. French Renaissance style cabinet, walnut veneered and blended. Rich carvings, gold plated fittings. Size 49¼ inches high, 48 inches wide, 25½ inches deep. Plays a complete program of approximately one hour's duration for dinner music, dancing and general entertainment.

The Automatic Orthophonic Victrola changes its own records. It plays twelve ten or twelve inch records without operating attention, at intervals of thirty seconds. Motor operating the mechanism stops automatically after the last record on magazine spindle has been played.

Operates from electric light socket on alternating current with induction disc motor, direct or alternating when equipped with Universal motor.

Three position index lever controls playing of ten or twelve inch records; or for playing individual records like other Orthophonic Victrolas.

For more complete information refer to page 22.

AUTOMATIC ORTHOPHONIC ELECTROLA
Number *Ten-Fifty-One*   *List Price*—$1050.00

In the same design cabinet we combine the automatic feature and the Electrola amplifying system, with the reproduction through the Orthophonic tone chamber.

For more complete information refer to page 22.

4

ORTHOPHONIC VICTROLA
Number *Eight-Thirty*

List Price—Spring Motor $330.00. Induction Disc Electric Motor $335.00
Universal Electric Motor $355.00

The Orthophonic Victrola in its highest development. Credence type cabinet walnut or mahogany veneered, blended finish, with Italian Renaissance decorations. Height 46″, width 31½″, depth 22″.

Orthophonic reproduction.
Non-set automatic eccentric groove brake. Record stops automatically without presetting.
Capacity for eighty records.
Spring motor runs twenty minutes without rewinding.

ORTHOPHONIC VICTROLA
Number *Eight-Thirty*

With front paneled in tooled leather.

List Price—Spring Motor $350.00. Induction Disc Electric Motor $385.00
Universal Electric Motor $405.00

Same equipment and design as the *Eight-Thirty* with delicately tinted tooled leather paneling, which blends admirably with the satin finish of the walnut cabinet.

5

VICTROLA
Number *One-One*

VICTROLA
Number *One-Two*

VICTROLA
Number *One-Seventy*

ORTHOPHONIC VICTROLA
Number *One-Ninety*

VICTROLA
Number *One-One*
List Price—$17.50

A semi-portable Victrola, cabinet finished in mahogany. Size 6½″ high, 12½″ wide, 13¾″ deep.

The lowest priced Victrola.
Table Type semi-portable.

Easy for the children to operate.
Plays all Victor Records.

VICTROLA
Number *One-Two*
List Price—$18.00

A semi-portable Victrola, cabinet finished in white enamel with brilliant colored decorations. Size 6½″ high, 12½″ wide, 13¾″ deep.

Plays all Victor Records. Sturdy little instrument made especially for children's playroom.
Easy for the children to operate.

VICTROLA
Number *One-Seventy*
List Price—$50.00

Cabinet mahogany veneered, blended finish, with maple overlay. Size 12½″ high, 17½″ wide, 14½″ deep.

Operated by spring motor.
Motor runs eight minutes without rewinding.
Equipped with Improved Victrola No. 4 Sound Box.
Semi-portable model.
May be moved from room to room.
Desirable for use in room with limited floor space since this model may be placed on convenient table or cabinet.

ORTHOPHONIC VICTROLA
Number *One-Ninety*
List Price—$75.00

Cabinet mahogany veneered, mahogany overlay—blended finish. Size 13½″ high, 19½″ wide, 18⅝″ deep.

Operated by spring motor.
Motor runs eight minutes without rewinding.
May be carried from room to room by any adult person.
An ideal instrument for use in room with limited floor space, since the instrument may be placed on convenient table or cabinet. Non-set automatic eccentric groove brake.
Record stops automatically without presetting.

20

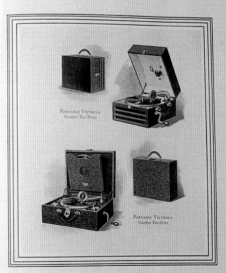

PORTABLE VICTROLA
Number *Two-Thirty*

PORTABLE VICTROLA
Number *Two-Sixty*

PORTABLE VICTROLA   Number *Two-Thirty*   *List Price*—$25.00

Cabinet finished in rich black crackle outside, brilliant mandarin red inside. Size 7¾″ high, 11¾″ wide, 14″ deep. Weight 17 pounds.

Concealed amplifying chamber.
Capacity for six ten-inch records carried in safety on the turntable when lid is closed.

Equipped with improved Victrola No. 4 sound box.
Plays all Victor records ten or twelve inch size.
Strong carrying handle, nickel fittings inside and out.

PORTABLE VICTROLA   Number *Two-Sixty*   *List Price*—$40.00

Encased in a durable leather-finished fabric. Choice of finish in dark blue with leather figured texture or brown with shark-skin texture. Size 7 inches high, 16½ inches wide, 13½ inches deep, weight 22 pounds.

Ingenious concealed amplifying chamber—the newest product of the Victor Research laboratories.
Built-in safety record container; holds twelve ten-inch Records. When the lid is raised the container automatically opens and is suspended between the lid and the turntable. Records easily accessible.

Equipped with improved Victrola No. 4 sound box.
Plays all Victor Records ten or twelve inch size.
Interior and exterior fittings gold-finished.
Genuine leather handle, flexible for comfortable carrying.
Spring clips inside the instrument for the removable winding key.

21

Opposite page: 5-39. The Columbia "Viva-tonal Portable" (Model "161") was introduced in 1928 for $50.00. Notice the prominent offset of the tone arm. The technological improvement in portable phonographs by the late 1920s was evident in this model. In fact, trade publications of the late 1920s, such as *Talking Machine World*, were literally awash in the tide of a portable phonograph craze. *Courtesy of Bob and Karyn Sitter.* (Value code: I)

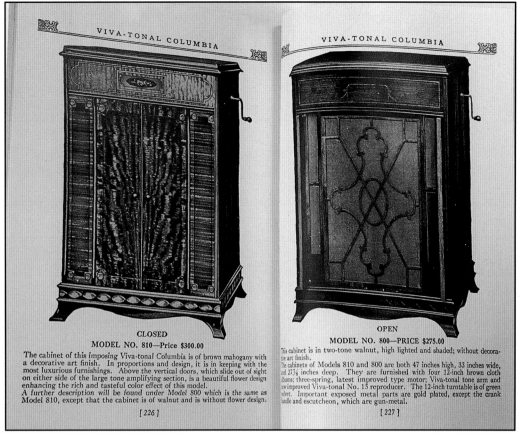

5-40. A Columbia 1927 catalogue that featured Models "810" ($300.00) and "800" ($275.00). The illustration on the right is especially telling. Despite the uncommon use of doors which retracted into the body of the cabinet, these Viva-tonal instruments were nearly identical copies of the "Orthophonic" Victor "Credenza."

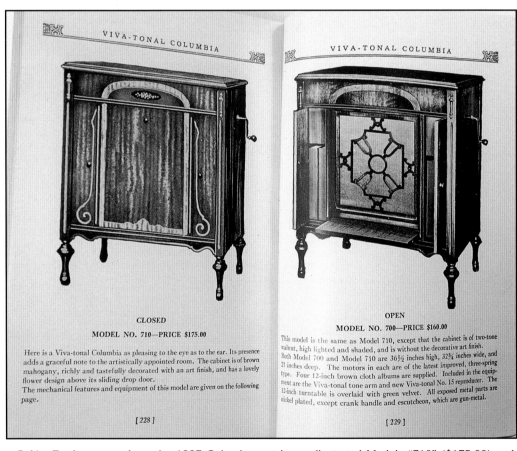

5-41. Further pages from the 1927 Columbia catalogue illustrated Models "710" ($175.00) and "700" ($160.00). Columbia added colored floral decorations to some of its Viva-tonal models, a conceit not exploited by other companies.

Victor's sales literally doubled from a shockingly low $20,857,955.76 in 1925 to $46,662,103.73 in 1926. Times were good, the stock market was soaring, and investments were the order of the day. Eldridge R. Johnson, founder of the Victor Talking Machine Company, was approached by the banking firms of J.&W. Seligman and Company and Speyer and Company to discuss a purchase of Victor. Johnson's clinical depression had made his recent absences at Camden longer and more frequent. Having weathered the talking machine climate for 30 years, he decided to sell. The Victor Talking Machine Company changed hands on December 7, 1926 for $40,000,000. Johnson took his share of the proceeds, $28,000,000, and joined the historic ranks of those who have encountered the tenuous link between wealth and happiness. Johnson's one great love was Victor. According to his son, he never quite filled the void that remained after the sale of his company.

TECHNOLOGY AND ART

The appearance of electrical recording and its reproduction either through exponential horns or vacuum tubes exerted a powerful influence on popular music. The talking machine had, of course, incited cultural fads such as a dance craze in the 'teens. Similarly, the proliferation and exchange of records had enabled (and would continue to enable) musicians to listen to their contemporaries, spreading new musical styles such as "jazz" around the world. Violinists had learned to incorporate vibrato into their playing in order to be "heard" by the acoustic recording horn. This stylistic necessity for recording gradually became expected on the concert stage as well, thus unexpectedly evolving instrumental technique. Electrical recording and improved reproduction nurtured new musical and vocal styles that could not be successfully recorded by the old acoustic method. Perhaps most noteworthy were the "whispering" tenors and baritones of the mid- to late-1920s. This singing style became recognized as "crooning," and remained popular for decades. Beyond all the critical acclaim for the artistry of Rudy Vallee and Bing Crosby, it was the electrical engineers who made it accessible for the masses. The talking machine industry had become a force of such power that improvements in technology were capable of spawning new stylistic art.

EDISON: "THE FIRST SHALL BE LAST"

Of the major talking machine firms of the 1920s, one stands out as an anachronism. Only Edison continued the cylinder format after 1922, despite steadily declining sales. The cylinder Amberola line consisted of the same models first offered in 1915: the table model Amberolas "30" and "50," and the floor model "75."

Surplus Edison Disc Phonograph cabinets were employed in the short-lived Amberola "60" of 1926, and the Amberola "80" of 1928. The cylinders themselves benefited from only one innovation of the 1920s: electrical recording. Unfortunately, by that time (November 1927), the cylinder business was so small that press runs of cylinder titles were 800 copies or less. Despite Thomas Edison's affection for the cylinder format, it represented a considerable financial drain on his company throughout the latter 1920s.

Edison's disc business was far more lucrative, but after the introduction of electrically-recorded lateral discs in 1925, the old vertically-cut acoustically-recorded Diamond Discs rapidly lost ground. Recording director Walter Miller wrote Edison a memo on January 14, 1926 in which he reported:

> From the various records I have listened to recently on the new Victor, which were electrically recorded, I am thoroughly convinced that we will sooner or later have to use electric recording if we wish to keep up with our competitors....Don't you think we could make some royalty arrangement with the A.T.& T. Co. for the use of this system? If not don't you think we ought to do some research work and try to do it ourselves?

Edison's response was painfully characteristic:

> Walter Miller = I do not want to touch this scheme at present. I could have taken this up without paying anybody. They cannot record without distortion. Edison.

Edison's business had always suffered from the lack of interchangeability between his products and those of his competitors. The comparatively high cost of Edison Diamond Disc Phonographs, notoriously noisy Edison record surfaces during the World War I years, and the lackluster record catalogue adversely affected sales prior to 1925. An August 27, 1923 memo from Walter Miller affirmed that, "All tunes that are recorded must be passed by Mr. Edison or the Committee at Orange." Additionally, Miller stated that, "All singers and instrumentalists used for recording have been passed by Mr. Edison, with the exception of two vaudeville artists." Thus a 76-year old functionally deaf man with no musical training was the principal judge of what went into the Edison record catalogue. After 1925, Edison products were perceived by many as obsolete, and for a time this had been true. Yet, "Edison Long Playing Records," playing 20 minutes on one 12" side, were introduced in November 1926. Electrical recording was offered in September 1927. Edison announced an "Edison Radio" in October 1928. In July 1929, the company introduced two portable phonographs designed to play lateral records, and the following month, the first Edison electrically-recorded lateral discs were unveiled.

All this might have saved the company had it occurred sooner, but the crucial years of 1920-1925 saw little or no real improvement in Edison products, and this lapse proved fatal. The "Long Play" project was rushed through in approximately 18 months. The record material proved susceptible to damage from the 2½ ounce weight exerted on the stylus. The delicate 450 threads-per-inch surfaces rapidly developed skips and repeats. The "Long Playing Records" were a failure. Edison electrical recording was a genuine improvement, but arrived two years late. Additionally, a new line of affordable Edison Phonographs designed to fully exploit the electrical discs was not developed. The "Edison Radio" was in actuality the re-named Splitdorf Radio, sub-

ject to the same flaws that characterized it when sold under the original brand. The Edison Portables were also subcontracted products, introduced eight years after Victor first offered the Victrola "50" portable. Finally, the superb Edison lateral cut records suffered the same fate of being introduced far too late to affect the ultimate success of the company. In October 1929, Thomas A. Edison, Incorporated ceased the production of entertainment Phonographs and records. Although the Ediphone Division would continue to offer office dictation machines for several decades, Edison was out of the music business. For the oldest name in recorded sound, time finally stood still.

5-42. In late 1927, Edison's youngest son, Theodore, began work on a system of recorded musical accompaniment for motion pictures. By January of 1928, the general design of the Ciné-Music apparatus had been established. A 14" condensite disc, similar to Edison's commercially-available Diamond Disc, played 80 minutes per side at 24 rpm and 360 threads-per-inch. It was at this precise moment that Vitaphone's synchronized sound-on-disc system was transforming the very nature of the movie industry (no connection with C.B. Repp's Vitaphone of 15 years earlier). Theodore Edison's Ciné-Music device, although electrically amplified, was non-synchronous and intended to assume a role that was about to disappear: that of the nickelodeon piano player. *Courtesy of the Edison National Historic Site.* (Value code: VR)

5-43. Despite the Edison company's material and intellectual resources, a lack of vision promoted by an ossified corporate structure crippled projects such as Ciné-Music with anachronistic responses to modern problems. The horn of the Ciné-Music prototype resembled an immense (41 3/4" diameter bell) version of the cygnet horn introduced by Edison two decades earlier. Whereas the Vitaphone system also employed an acoustic horn to amplify the sound, Vitaphones's horn was of the Western Electric (folded) "exponential" type, which had revolutionized the phonograph industry during the latter 1920s. Shown is the Ciné-Music system was outmoded at its inception and never commercially produced. *Courtesy of the Edison National Historic Site.*

Opposite page, top: 5-44. The Psycho-Phone was not an entertainment device, but one designed to play a pre-recorded message to its slumbering user during the night. Self-improvement through subliminal stimulation, post-hypnotic suggestion, and Freudian theory were popular topics in 1927, when the Psycho-Phone trademark was registered. The machine's inventor, Alois Saliger, designed the electrically-driven turntable to be triggered by a clock, much like an alarm. The soundbox was positioned at the start of a specially-recorded disc before the user retired to bed. When the clock started the motor, the sleeping individual, rather than waking, would hear a message played softly through the Psycho-Phone's tiny horn. At the end of the record, a switch below the motor board would turn off the electric motor. A few Psycho-Phone records survive, the contents of which can be represented by the disc labeled "Prosperity." The record intones: "Money wants me and comes to me. Business wants me and comes to me…I am rich; I am success…" *Courtesy of Douglas DeFeis.* (Value code: VR)

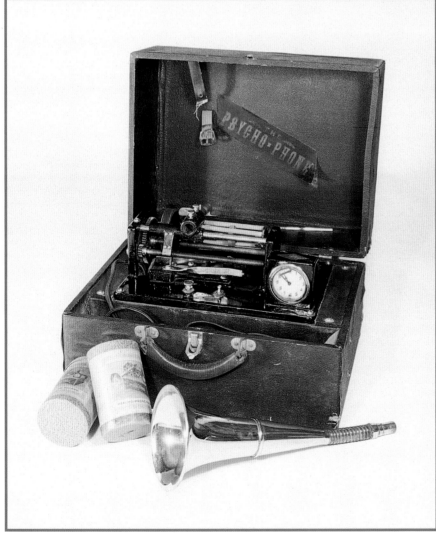

5-45. A good example of history repeating itself is represented in the cylinder incarnation of the Psycho-Phone. Like the cylinder talking machines of the early 1890s, the Psycho-Phone was an acoustic machine driven by an electric motor, surely the last of its kind for home use. The cylinder format allowed the user to record his own messages tailored to personal life struggles, such as, "Writing books makes me rich." (If only repetition could guarantee effectiveness!) These custom messages were subsequently played back at a preset time during sleep, to presumably imbed themselves in the subconscious mind of the subject. The cylinder Psycho-Phone was a remarkably robust mechanism, capable of being easily transported in its carrying case, along with a nickeled horn, support crane, recorder, reproducer, and cylinders. *Courtesy of David Giovanonni.* (Value code: VR)

5-46. Thorens was a Swiss music box and talking machine maker with a very venerable history in mechanical music. The company's popular "Excelda" portable disc machine, which imitated the shape of a Kodak "Autographic" camera, remained in production for decades. The first models, from the 1920s, used soundboxes with mica diaphragms. The later versions employed metal diaphragms, but changes to the basic design were minor. The "Excelda" came in a variety of colors including red, blue, green, tan, gray, and black. *Courtesy of Walter and Carol Myers.* (Value code: I)

5-47. This charming little table model with inlaid doors was constructed of Thorens parts. The Swiss, having pioneered in the manufacture of music boxes, produced an amazing variety of talking machines and related components throughout the first three decades of the twentieth century. *Courtesy of Lou Caruso.* (Value code: I)

5-48. The Belknap circus wagon was a charming musical toy made by the Charles H. Belknap Company of Brooklyn, New York. Made of wood, brightly painted and decaled, it sported a miniature cardboard band. Note the crank protruding from the front of the device, which hints at the truck's musical secret. *Courtesy of Walter and Carol Myers.* (Value code: VR)

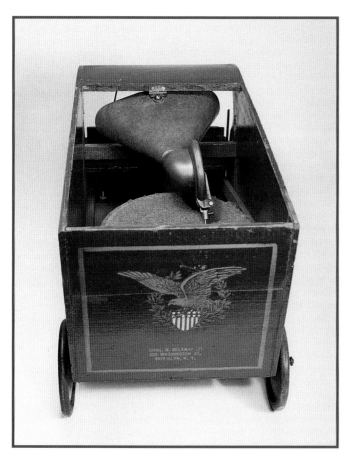

DIRECTIONS

Place handle through hole in dash. Screw it into winding socket.

Put men on seats. (Liquid glue can be used to fasten them when found necessary.)

Put needle in sound producer and place record on turn table.

Care should be exercised in not over-winding spring.

If the wagon is to travel while being played, screw up small brass nut on right rear wheel.

Should wagon not start promptly, the turn table can be started or the wagon given a gentle shove.

To stop motor, use brake.

A large assortment of records can be obtained at your own dealer's, or any 5c and 10c store.

MANUFACTURED BY
CHAS. H. BELKNAP CO.
BROOKLYN, N. Y.

5-50. The instructions for the circus wagon were pasted to the underside of the roof. The reference to easily obtainable small diameter discs also suggests the late twenties, when 7" diameter children's records such as "Cameo Kid" and "Playtime" were readily available. *Courtesy of Walter and Carol Myers.*

5-49. Inside the circus truck was a small acoustic disc talking machine mechanism of the late-twenties "Genola" style. The sound emerged from behind the driver. The phonograph motor also could be engaged to drive the wagon forward as the record played. *Courtesy of Walter and Carol Myers.*

5-51. This hybrid Phonograph was created for the personal use of Thomas Edison at his laboratory in Orange, New Jersey. The (Herzog-made) cabinet was salvaged from an Amberola "IA" cylinder Phonograph (1909-1910). The mechanism was that of a full-sized "Laboratory" model Diamond Disc Phonograph (circa 1915). Edison's impaired hearing, at this point in his life, made it necessary for him to sit directly in front of a horn in order to audition records. For this reason, the grille of this custom-adapted machine was discarded from the front of the cabinet, and a folding writing desk substituted, on which Mr. Edison could lean and make notes. A paper tag accompanying this machine is hand-written by Theodore Edison's wife as follows: "This is a special phonograph for Mr. Edison's use. Please do not move without permission from him. 8/21/1928, Mrs. T. M. Edison." *Courtesy of the Edison National Historic Site.* (Value code: VR)

5-52. In early 1928, Edison began manufacturing this small Phonograph mechanism to fit inside cigarette vending machines operated by the Automatic Merchandising Corporation of America. Each 7 1/4" diameter disc held four separate two-second recordings. The reproduction was acoustic, and each message was limited to "thank you" and a short slogan. Production continued on a limited basis until the Phonograph Division closed in the fall of 1929. *Courtesy of the Edison National Historic Site.* (Value code: VR)

5-53. The reproducer from this particular "slogan" machine was an interesting adaptation. An Edison "needle cut" soundbox was fitted with a modified needle bar and diamond stylus to play the vertically-recorded cigarette machine discs. *Courtesy of the Edison National Historic Site.*

5-54. In September 1928, Edison introduced its first combination Radio-Phonographs. Two models, the "C-1" and the "C-2" (shown) were offered, followed by the "C-4" a year later. The "C-2" was priced at $495.00 (less tubes). The two front doors opened flat against the sides of the cabinet, allowing the four record albums to be easily removed. *Courtesy of Robin and Joan Rolfs.* (Value code: G)

5-55. The playing compartment of the Edison "C-2" Radio-Phonograph featured an electric pickup capable of playing lateral and vertical cut records. This component was designed by T. A. Edison's youngest son, Theodore. A September 1928 advertisement extolled, "The cabinet of blended walnut finish, with ornamental panels of burl maple, fits pleasingly into any interior, harmonizing with other furniture." The "C-2" measured 28 1/4" wide x 18" deep x 48 1/2" tall. *Courtesy of Robin and Joan Rolfs.*

5-56. This advertising sign touts the New Edison's ability to play both vertical and lateral cut discs. Unfortunately, this feature arrived a decade too late. *Courtesy of the Charles Hummel collections.*

5-57. As the popularity of radio and electrically recorded discs grew, several companies offered electric turntables to play records through existing radios. A precursor to modern component systems, these units would be plugged into the wall and connected by a cable to the radio. Shown is a Polk "Radiotrola," sold by James K. Polk Incorporated of Atlanta, Georgia. The electrical pickup was a Pacent "Phonovox," manufactured by a New York City firm. The cabinet was styled to resemble a radio of the period when the lid was closed. (Value code: I)

5-58. The manufacture of portable talking machines throughout the 1920s had proved lucrative for many companies. Edison, clinging to the vertical cut Diamond Disc and its necessary feed mechanism, was prevented from sharing in the portable machine business until too late to benefit from it. In July 1929, the Edison Needle Type Portable Phonographs were introduced. Two models were offered, called the "P-1" (shown) and the "P-2." The most noticeable difference between the two was the gold-plated hardware of the "P-1." Note, the soundbox of the "P-1" was not finished in gold. Priced at $35.00, the "P-1" was advertised as "The Little Portable with the Big Console Tone." Both catalogued Edison Portables seem to have been manufactured by the Prime Manufacturing Company of Milwaukee, Wisconsin. *Courtesy of Robin and Joan Rolfs.* (Value code: G)

5-59. The smaller of the two Edison Portables was the "P-2." Originally intended to sell for $25.00 retail, this particular example retains its "sale" price: probably as a result of Edison's discontinuation of phonograph and record manufacture in November 1929. *Courtesy of Martin F. Bryan.* (Value code: H)

5-60. When Edison introduced its new "Needle Cut" records in July 1929, certain classical recordings were available as 12" diameter discs. These, and the 12" Long Play discs of 1926, were the only records of this size that Edison offered to the public. *Courtesy of Martin F. Bryan.*

5-61. This uncatalogued Edison portable is from the "needle cut" period (1929). The machine bore similarities to Columbia Viva-tonal portables "No.161" and "No.163." The documented Edison portables ("P-1" and "P-2") differed considerably, suggesting that this machine was contracted to a different manufacturer. Modern collectors have asked why a company of Edison's stature would have put its efforts into marketing innocuous portables. This has been viewed as a sign of Thomas A. Edison, Incorporated's failing judgement. However, when seen in the context of a very well documented portable craze, which gripped the late 1920s, the Edison portables take on new, and well-deserved, meaning. *Courtesy of the Edison National Historic Site.* (Value code: VR)

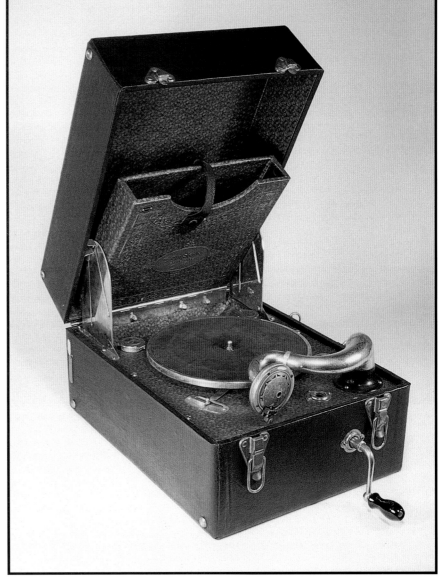

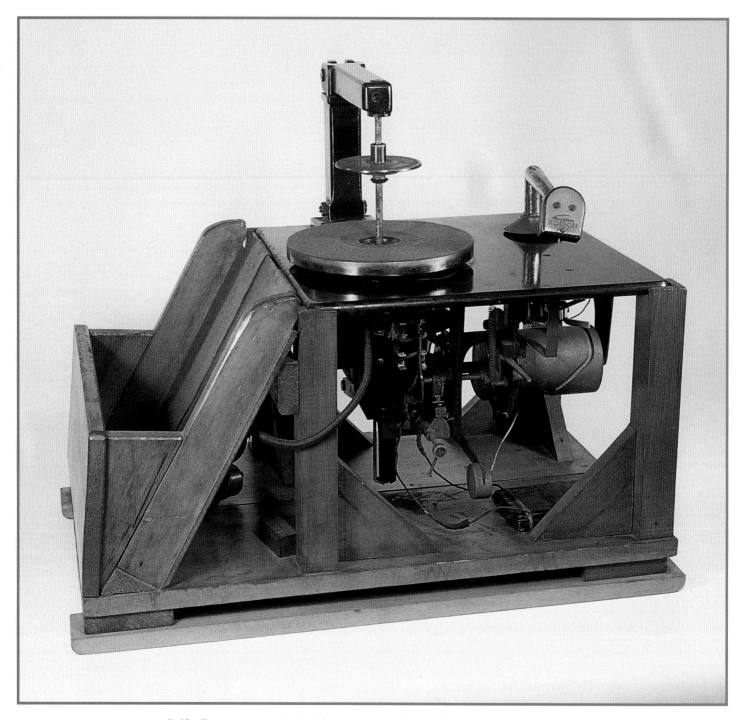

5-62. For a company that had lost its way by obstinately offering outmoded products for such a long time, 1929 was a remarkably fertile year for the Phonograph Division of Thomas A. Edison, Incorporated. This prototype device represented one of several projects under development at that time. The automatic changer mechanism incorporated electric amplification and drive. Similar to certain Victor models, the bin at the left caught 78s after they had been played and ejected. *Courtesy of the Edison National Historic Site.* (Value code: VR)

5-63. The Brunswick-Balke-Collender Company began electrical recording in April 1925, issuing the first of its new process discs in August. Victor and Columbia were preparing to unveil their new re-entrant horn acoustic machines: the "Orthophonic" (November 1925) and the "Viva-tonal" (January 1926). Brunswick trumped them both by introducing its all electric "Panatrope" on November 11, 1925. This was the first electric turntable/electrically amplified talking machine for home use. The Model "3KR6" Brunswick Panatrope with Radiola (shown) was introduced in February 1929. *Courtesy of Lou Caruso.* (Value code: G)

5-64. The Brunswick Model "3KR6" Panatrope with Radiola in the opened position. After revolutionizing the talking machine industry with its all electric Panatrope, Brunswick curiously introduced a line of acoustic, non-electric Panatropes in April 1926. In May and June 1927, Brunswick advertised its acoustic Panatropes as "Prismatones," but by July had reverted to the "Brunswick Panatrope, Exponential Type." However, the all-electric Panatropes and Panatrope-Radiola combinations continued to be the flagships of the Brunswick line. *Courtesy of Lou Caruso.*

5-65. The interior of the Model "3KR6" Panatrope with Radiola. Note the unusual design of the tone arm and the pronounced offset of the electrical pickup. Like other instruments of the period, the cabinet interior presents a capacious appearance. *Courtesy of Lou Caruso.*

5-66. There is something sublimely absurd about the idea of releasing a brand new style of Edison cylinder Phonograph in the fall of 1928. Had there been a prayer of a chance that anyone would care about the introduction of the Edison Amberola "80," the model might have been termed "ill-timed" or "financially unfeasible." Since only a small number of people even remembered that the cylinder Phonograph continued to exist, the inauguration of the Amberola "80" easily transcended commercial considerations altogether. It was a crazy act of devotion, like leaping onto your spouse's burning funeral pyre. The "80" was the first new style of floor-standing Amberola in 13 years. The mechanism was basically that of the old "75," though an improvement was made to the governor, and the reproducer was fitted with a heavier floating weight (the "Diamond D"). The cabinet was adapted from the Diamond Disc "S-19." At first glance, the two machines looked identical, except for the position of the crank hole. The cabinet hardware was gold-plated, for the first time in the Amberola line since the demise of Amberolas "IB" and "III" over 13 years earlier. Inside the bottom cupboard of the "80" were two trays which held cylinders. By the fall of 1929, when Thomas A. Edison, Incorporated exhausted the limits of devotion and ceased entertainment Phonograph production, only a small number of Amberola "80"s had been sold. (Value code: F)

The unique ability of the talking machine to make time stand still—to encapsulate, preserve and reproduce sound—attracts and charms us today. Through those traits, the talking machine has become an anthropological tool of ever-increasing importance. The past continues to recede from our grasp, but by lowering a stylus or needle to a record, the aural history it preserves is made new again. A listener today can hear voices of the 1870s and 1880s intoning utterances such as "mama," "papa," "one o'clock, two o'clock," and "I hope you haven't got such a bad cold as I have." John Phillip Sousa conducts the Marine Band. Joseph Jefferson recites from *Rip Van Winkle.* A young Caruso sings again. An aging Adelina Patti performs from the repertoire that she made legendary. Theodore Roosevelt speaks about the nation he loves. General Pershing affirms that victory is inevitable. Thomas Edison encourages international respect. Paul Whiteman's Orchestra plays. Bessie Smith belts out her songs. Robert Johnson picks Southern blues on his guitar. George Gershwin renders seminal compositions that define a generation. The talking machine brings them to life, if only for a few minutes.

The "talking" machine. What other mechanical device has been euphemized with such a uniquely human characteristic? From its beginnings, the talking machine was seen as something special, almost supernatural.

Despite its usually straightforward mechanics and surprisingly simple acoustic principles, the early talking machine continues to suggest an unseen presence, a magical door that defies death and the passage of time. There were many accounts in the 1890s and early 1900s of talking machines playing recordings of voices whose owners had already passed away. These "voices from the grave" were a new experience for a society where a voice had never been heard outside the presence of the speaker.

A century later, despite the vast recorded resources amassed around the world, we still long for the voice of Abraham Lincoln, Jenny Lind, Walt Whitman or Mark Twain. It seems tragic that the spirits of Lincoln and Lind are irretrievable; that their voices were never captured on wax; and that recordings of Whitman and Twain apparently have been lost. Yet the advent of the talking machine immortalized thousands of others. It may be merely a snatch of recorded speech, or an inspiring musical legacy. The spirit may whisper to you as though from another room, or reveal its virtuosity with stunning presence. It lives in a polished wooden cabinet, and sings through a shining horn. For the people who shaped the history of early recorded sound, the talking machine is a monument fit for the ages.

Value Code Key

Preparing a value system for a number of largely unusual talking machines is a task fraught with concerns. Of the systems commonly in use, we have adopted the "Letter Designated Grading Guide." Needless to say, everyone has his own special axe to grind when it comes to price guides. Since the publication of our book *The Talking Machine, an Illustrated Compendium 1877-1929*, we have been treated to every sort of theory. Some demanded that the prices be in specific dollar amounts and higher than retail. Some claimed values should be much lower than retail. Others insisted upon a range. No one will be completely satisfied, and some could be agitated by our choice of the letter system. We stand behind our decision. In keeping with precedent, we will designate prices *only* for talking machines, not peripheral items. Items that are one-of-a-kind or uncommonly sold will be rated "very rare" (VR). Since machines of this caliber constitute a primary theme of this book, there will be a good number of "VR"s.

PLEASE TAKE NOTE, BEFORE YOU USE THIS GUIDE: The "VR" (very rare) designation DOES NOT indicate a price in excess of price code "A." The two categories are unrelated in terms of dollars. "VR" is intended to distinguish machines that are not commonly traded. Although "VR" machines might sell for a wide variety of prices, the reason that they are not valued is because they change hands so infrequently.

Values will be expressed *in the individual captions* by the following letter codes:

A	$12,000. to $15,000.
B	$9,000. to $12,000.
C	$6,000. to $9,000.
D	$4,000. to $6,000.
E	$2,500. to $4,000.
F	$1,500. to $2,500.
G	$750. to $1500.
H	$350 to $750.
I	under $350.

In using the value codes, we advise *conservatism*. If you are trying to establish the value of an object in your possession, the natural instinct is to *seize* upon the highest price in the given range, regardless of condition. This can be a mistake. The condition of the items in the book is nearly always excellent, but the condition of *your* item is the important issue. Condition can affect the value by up to 50%—and occasionally more. The bottom end of a price range has as much validity as the top, though it seems a less attractive prospect. Different examples of the same type of object may represent the entire range of a certain price category, due to condition. The ranges are also intended to be *comparative*. Certain objects in the "H" category, for instance, may *never* reach $750.00 no matter how perfectly preserved they are.

Beyond condition, many, many factors affect the price of an "antique." Exact worth is something that can not be fixed precisely in time. We can only hint at it. Our guide is intended to help you understand the value of the objects illustrated, but we vigorously discourage the use of the phrase, "It *books* for such-and-such a price." True worth is what the market will bear. Quoting books, auction results or what a shop is asking will probably not elevate the price higher than the general market will currently support.

Bibliography

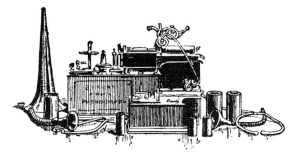

Association for Recorded Sound Collections, various ed.s: *ARSC Journal*, various issues to present.

Barnum, Frederick O. III: *His Master's Voice in America*, Camden, General Electric Company, 1991.

Barr, Steven C.: *The Almost Complete 78 RPM Record Dating Guide (II)*, Huntington Beach, Yesterday Once Again, 1992.

Baumbach, Robert W.: *Columbia Phonograph Companion, Vol. II (Disc Graphophones and the Grafonola)*, Woodland Hills, Stationery X-Press, 1996.

Baumbach, Robert W.: *Look for the Dog*, Woodland Hills, Stationery X-Press, 1996.

Bayly, E., ed.: *The EMI Collection*, Bournemouth (GB), Talking Machine Review, 1977.

Bayly, E., ed: *The Talking Machine Review International*, Bournemouth (GB), various issues beginning 1970.

Bill, Edward Lyman, ed.: *The Talking Machine World*, New York, various issues 1905-1929.

Bryan, Martin, ed.: *The New Amberola Graphic*, St. Johnsbury (VT), various issues 1973 to present.

Charosh, Paul, compiled by: *Berliner Gramophone Records, American Issues 1892-1900*, Westport, Greenwood Press, 1995.

City of London Phonograph and Gramophone Society, various ed.s: *The Hillandale News*, various issues 1960s to present.

Conot, Robert: *A Streak of Luck*, New York, Seaview Books, 1979.

Dethlefson, Ronald, ed.: *Edison Blue Amberol Recordings, 1912-1914*, Brooklyn, APM Press, 1980.

Dethlefson, Ronald, ed.: *Edison Blue Amberol Recordings, 1912-1914 (paper cover)*, Woodland Hills (CA), Stationery X-Press, 1997.

Dethlefson, Ronald, ed.: *Edison Blue Amberol Recordings, 1915-1929*, Brooklyn, APM Press, 1981.

Dethlefson, Ronald, ed.: *Edison Blue Amberol Recordings, 1915-1929*, second edition, Los Angeles, Mulholland Press, 1999.

Edge, Ruth and Petts, Leonard: *The Collector's Guide to 'His Master's Voice' Nipper Souvenirs*, London, EMI Group, 1997.

Fabrizio, Timothy C. and Paul, George F.: *Antique Phonograph Gadgets, Gizmos and Gimmicks*, Atglen (PA), Schiffer Publishing, Ltd., 1999.

Fabrizio, Timothy C. and Paul, George F.: *The Talking Machine, an Illustrated Compendium 1877-1929*, Atglen (PA), Schiffer Publishing, Ltd., 1997.

Fagan, Ted, compiled by: *The Encyclopedic Discography of Victor Recordings, Pre-Matrix Series*, Westport, Greenwood Press, 1983.

Fagan, Ted, compiled by: *The Encyclopedic Discography of Victor Recordings*, Westport, Greenwood Press, 1986.

Frow, George L.: *Edison Cylinder Phonograph Companion*, Woodland Hills, Stationery X-Press, 1994.

Frow, George L.: *The Edison Disc Phonographs and the Diamond Discs*, Sevenoaks, Kent (GB), George L. Frow, 1982.

Gaisberg, Frederick W.: *The Music Goes Round*, North Stratford (NH), Ayer Company Publishers, Inc., 1977 (1942 reprint).

Gelatt, Roland: *The Fabulous Phonograph, revised*, New York, Appleton-Century, 1965 (three ed.s 1955-1977).

Gracyk, Tim, ed.: *Victrola and 78 Journal*, Roseville (CA), various issues 1994-1998.

Hatcher, Danny, ed.: *Proceedings of the 1890 Convention of Local Phonograph Companies*, Nashville, Country Music Foundation Press, reprint, 1974.

Hazelcorn, Howard: *A Collector's Guide to the Columbia Spring-Wound Cylinder Graphophone, 1894-1910*, Brooklyn, APM Press, 1976.

Hazelcorn, Howard: *Columbia Phonograph Companion, Volume I: Hazelcorn's Guide to the Columbia Cylinder Graphophone*, Los Angeles, Mulholland Press, 1999.

Hunting, Russell, ed.: *The Phonoscope*, New York, Phonoscope Publishing Company, various issues 1896-1900.

Johnson, E. R. Fenimore: *His Master's Voice Was Eldridge R. Johnson*, Milford (DE), State Media, Inc., 1975.

Koenigsberg, Allen, ed.: *The Antique Phonograph Monthly*, Brooklyn, APM Press, various issues 1972-1993.

Koenigsberg, Allen: *Edison Cylinder Records, 1889-1912, 2nd ed.*, Brooklyn, APM Press, 1988.

Koenigsberg, Allen: *The Patent History of the Phonograph, 1877-1912*, Brooklyn, APM Press, 1991.

Lorenz, Kenneth M.: *Two-Minute... Cylinders of the Columbia Phonograph Company*, Wilmington (DE), Kastlemusick, Inc., 1981.

Maken, Neil: *Hand-Cranked Phonographs*, Huntington Beach, Promar Publishing, 1995.

Marco, Guy A. and Andrews, Frank, eds.: *Encyclopedia of Recorded Sound in the United States*, New York, Garland Publishing, Inc., 1993.

Martland, Peter: *Since Records Began: EMI: the First Hundred Years*, (distributed by Timber Press, Portland, OR), Amadeus Press, 1997.

Marty, Daniel: *Histoire Illustrée du Phonographe (Illustrated History of the Phonograph)*, Lausanne/Paris, Edita/Lazarus, 1979, (reprinted in various English language editions with slight modifications of the title).

Michigan Antique Phonograph Society, various ed.s.: *In the Groove*, various issues to present.

Moore, Jerrold Northrop: *A Matter of Records*, New York, Taplinger Publishing Company, 1977.

Moore, Wendell, ed.: *The Edison Phonograph Monthly*, various anthologies 1903-1916, New Albany (IN), Wendell Moore Publications.

Proudfoot, Christopher: *Collecting Phonographs and Gramophones*, New York, Mayflower Books, 1980.

Read, Oliver and Welch, Walter L.: *From Tinfoil to Stereo*, Indianapolis, Howard W. Sams and Company, Inc., 1959.

Reiss, Eric L.: *The Compleat Talking Machine*, Chandler (AZ), Sonoran Publishing, (revised)1998.

Rust, Brian, compiled by: *Discography of Historical Records on Cylinders and 78s*, Westport, Greenwood Press, 1979.

Sherman, Michael W. (with Moran, William R. and Nauck, Kurt R. III): *Collector's Guide to Victor Records*, Dallas, Monarch Record Enterprises, 1992.

Sherman, Michael W. and Nauck, Kurt R. III: *Note the Notes, An Illustrated History of the Columbia 78 rpm Record Label, 1901-1958*, Monarch Record Enterprises, New Orleans, 1998.

Sutton, Allan: *Directory of American Disc Record Brands and Manufacturers, 1891-1943*, Westport, Greenwood Press, 1994.

Welch, Walter L. and Burt, Leah Brodbeck Stenzel: *From Tinfoil to Stereo, 1877-1929*, Gainesville, University Press of Florida, 1994.

Wile, Raymond R. and Dethlefson, Ronald: *Edison Diamond Disc Recreations, Records and Artists 1910-1929*, Brooklyn, APM Press, 1985.

Glossary

AMBEROL: The first four-minute cylinder developed by Edison, 1908-1912. These cylinders are black in color, made of especially hard metallic soap, and were usually sold in green and white containers.

AMBEROLA: The name denoting Edison's line of internal-horn cylinder Phonographs, 1909-1929.

BACK MOUNT: A term used to describe a talking machine which uses a back bracket which supports the horn and thus removes its inertia from the soundbox.

BEDPLATE: The plate (usually metal) to which the upper mechanism of a cylinder talking machine is mounted.

BLANK: A smooth, grooveless unrecorded cylinder.

BLUE AMBEROL: The blue celluloid four-minute cylinder sold by Edison, 1912-1929. The containers for these cylinders were blue until 1917, orange and blue thereafter.

CARRIAGE: The assembly which "carries" the reproducer of a cylinder talking machine across the recording.

CARRIER ARM: Edison nomenclature for the carriage. The half-nut and spring are attached directly to the carrier arm on nearly all Edison machines.

COMBINATION ATTACHMENT: Devices offered by Edison to convert pre-1908 Phonographs to play four-minute cylinders in addition to the two-minute variety.

CONCERT: Edison's trade name for the five-inch diameter cylinder and the Edison Phonograph designed to play it.

CRANE: The support used to mount horns of 15-inch length or greater to cylinder talking machines.

CYLINDER: A geometric form on which entertainment recordings were made, 1877-1929. The earliest sheets of tinfoil were followed by self-supporting cylinders of ozocerite-covered cardboard, stearic acid/paraffin, hard metallic soap, and celluloid. Advantages of the cylinder format included a constant surface speed and the ability to make home recordings.

DIAMOND DISC: Edison's line of disc records and machines, 1912-1929. The Diamond Discs were vertically recorded, and the Diamond Disc Phonographs were one of the very few disc talking machines to use a feedscrew.

DIAPHRAGM: The flexible, circular vibrating membrane of a soundbox or reproducer which converts mechanical energy to acoustic energy or sound waves. Mica, glass, compressed paper, copper and aluminum were commonly used as diaphragms in early talking machines.

ELBOW: The connection between the horn and the soundbox or tone arm of a disc talking machine. Earliest elbows are made of leather, gradually giving way to metal elbows after 1900.

ENDGATE: A swinging arm which supports one end of the mandrel on some cylinder talking machines. To mount or remove a cylinder it is therefore necessary to open and close the endgate.

FEEDSCREW: A threaded rod which usually drives a half-nut fixed to the carriage of a cylinder talking machine, thus guiding the reproducer across the record grooves. Certain disc machines used feedscrews to drive the soundbox across the record or to move the turntable beneath a fixed soundbox. Similarly, some cylinder machines used feedscrews to drive a mandrel longitudinally beneath a fixed reproducer.

FRONT MOUNT: A term used to describe a disc machine where the horn attaches directly to the soundbox, and the support arm (or mount) runs directly below and parallel to the horn. In such an arrangement, the support arm points in the same general direction as the bell of the horn.

GRAMOPHONE: Originally the name by which Emile Berliner's disc talking machine was known, it became a generic term to denote any disc playing talking machine, but fell out of use in the United States.

GRAND: Columbia's trade name for the five-inch diameter cylinder and the various Columbia machines designed to play it.

GOVERNOR: The mechanical assembly in a talking machine motor which regulates the speed, usually by limiting the outward movement of spinning weights.

HALF-NUT: An internally threaded metal piece usually in the form of a nut which has been cut in half. The threads of the half-nut correspond to the threads of the feedscrew. As the feedscrew revolves, the half-nut will be propelled along it, thus driving the carriage of a talking machine.

LATERAL RECORDING: The type of disc talking machine recording where the information is encoded in sides of the groove. (see VERTICAL RECORDING)

MANDREL: The tapered drum upon which the cylinder record is placed for playing.

NEEDLE: The point (usually steel) of the soundbox which rides the grooves of the recording and transmits vibrations to the diaphragm.

NEEDLE TIN: A container to store and dispense steel talking machine needles.

NEEDLE SHARPENER (OR CUTTER): A device for repointing disc talking machine needles of fibre, thorn or steel.

RECORDER: A device comprising a diaphragm and cutter stylus for recording cylinder or disc records.

REPEATING ATTACHMENT: A device for cylinder or disc talking machines which returns the needle or stylus to the beginning of the record after the selection has played.

REPRODUCER: The component comprising stylus, linkage and diaphragm, which reproduces the sound from the record grooves. This term is most frequently applied to cylinder machines. (See SOUNDBOX)

SHAVER: On a cylinder talking machine, a device which holds a knife in position to pare the recorded surface from a wax cylinder.

SOUNDBOX: The component comprising needle, linkage and diaphragm, which reproduces the sound from the record grooves. This term is most frequently applied to disc machines. (See REPRODUCER)

STYLUS: The point (usually sapphire or diamond) of the reproducer which rides the grooves of the recording and transmits vibrations to the diaphragm.

TONE ARM: A movable hollow tube which conducts sound to the horn from the soundbox.

TRUNNION: Rightly, the sleeve visible on either side of the carriage which slides over the feedscrew shaft on a Graphophone, but commonly used to denote the entire carriage.

VERTICAL RECORDING: The type of cylinder or disc recording in which the sound vibrations are encoded in the bottom of the groove. (see LATERAL RECORDING)

WAX: Although the material from which many cylinder records were made is generally referred to as "wax," it was more rightly various formulations of metallic soap. The resemblance of this material to conventional wax promotes the common descriptive term.

Index